Beginning Algebra I & II

M. Gaul, B. Goldner, E. Jasso,
D. Li, P. Lippert, E. Murphy

North Seattle College

Second Edition
Winter 2017

TABLE OF CONTENTS

Mathematics instructors from North Seattle College (NSC) created this workbook to better serve our students and to align our curriculum with NSC learning outcomes. This workbook assumes no prior knowledge of algebra, but we do assume students are familiar with the basic rules of arithmetic. There is an Appendix covering these topics for those students that may need a review, or to be used as reference along with the book.

This workbook is organized as follows:

Chapters

- **Sections** are the main instructional component for each chapter. They are where ideas are introduced.

- **Worked Examples** provide further explanation of the concept. It is recommended that students read through these examples carefully.

- **Class/Media Examples** can be used as classroom examples by instructors.

- **You Try** are problems embedded in the workbook to help students reinforce the concepts. It is recommended to solve all these problems an to do so in the order they appear, showing as much work as possible in a neat and organized fashion. There is space provided to work the solutions in the workbook.

Practice Problems

- There is a set of **Practice Problems** at the end of each section. The only way to learn math is by practice. We suggest that students attempt them on their own first before seeking help. Although this is a workbook, and there is space to show most of the work, we recommended that students keep an organized notebook for this class. Students should do the practice problems in this notebook, showing all the worked solutions in a neat and organized way, so they can refer to them easily.

- Solutions to all the Practice Problems have been compiled and each instructor can make them available to the students.

CHAPTER 1

LINEAR EQUATIONS

1.1. ORDER OF OPERATIONS

Objective: Evaluate expressions using the order of operations, including the use of absolute value.

When simplifying expressions it is important that we simplify them in the correct order.

Order of Operations:

Parenthesis (Grouping Symbols)
Exponents
Multiply and Divide (Left to Right)
Add and Subtract (Left to Right)

Multiply and Divide are on the same level because they are the same operation (division is just multiplying by the reciprocal). This means they must be done left to right, so some problems we will divide first, others we will multiply first. The same is true for adding and subtracting (subtracting is adding the opposite).

Often students use the word PEMDAS to remember the order of operations, as the first letter of each operation creates the word PEMDAS. If we think about PEMDAS as a vertical

word written as: $\begin{matrix} P \\ E \\ MD \\ AS \end{matrix}$ we are more likey t o remember that multiplication and division are

done left to right (same with addition and subtraction). Another way students remember the order of operations is to think of a phrase such as *"Please Excuse My Dear Aunt Sally"* where each word starts with the same letters as the order of operations start with.

| **Problem 1 : | *Worked Example* |

Simplify: $2+3(9-4)^2$

Solution.

$$2+3(9-4)^2 \quad \text{Perform operation inside parenthesis first}$$
$$=2+3\,(5)^2 \quad \text{Apply exponents}$$
$$=2+3(25) \quad \text{Multiply}$$
$$=2+75 \quad \text{Add}$$
$$=77 \quad \text{Our Solution}$$

In the previous example, if we had added first, 125 would have been the answer, which is incorrect.

Problem 2 : | *Worked Example*

Simplify: $30 \div 3(-2)$

Solution.

$$
\begin{aligned}
\underbrace{30 \div 3}(-2) \quad &\text{Divide first (left to right!)}\\
=\underbrace{10(-2)} \quad &\text{Multiply}\\
=-20 \quad &\text{Our Solution}
\end{aligned}
$$

In the example above, if we had multiplied first, the answer would have been -5, which is incorrect.

The example below illustrates an important point about exponents. Exponents only are considered to be on the number they are attached to. This means when we see -8^2, only the 8 is squared giving us $-(8^2) = -(8 \cdot 8) = -64$. But when the negative is in parenthesis, such as $(-8)^2$, the negative is part of the number and is also squared. That means, $(-8)^2 = (-8)(-8) = 64$. $(-8)^2$ gives us a positive soultion, 64, whereas -8^2 gives us a negative solution, -64.

Problem 3 : | *Worked Example*

Simplify: $-4^2 - (-5)^2$

Solution.

$$
\begin{aligned}
-4^2 - (-5)^2 \quad &\text{Rewrite } 4^2 \text{ as } 4 \cdot 4 \text{ and } (-5)^2 \text{ as } (-5 \cdot -5)\\
=-\underbrace{(4 \cdot 4)} - \underbrace{(-5 \cdot -5)} \quad &\text{Perform Parentheses}\\
=\underbrace{-16 \quad - \quad 25} \quad &\text{Subtract}\\
=-41 \quad &\text{Our Solution}
\end{aligned}
$$

If there are several parenthesis in a problem we will start with the inner-most parenthesis and work our way out. Inside each parenthesis we simplify using the order of operations as well. To make it easier to know which parenthesis goes with which parenthesis, different types of parenthesis will be used such as { }, [ ] and ( ). These all mean the same thing. They are parenthesis and must be evaluated first.

Problem 4 : | *Media/Class Example*

Simplify the following.

a) $17-3\cdot5$

d) $\frac{2}{5}(1+4\cdot3)$

b) $3+2(4-15)$

e) $(9-3)\div(5+17)$

c) $5-3(2-4^2)$

f) $5-3(2-4)^2$

It can take several steps to complete a problem. The key to successfully complete an order of operations problem is to take the time to show your work clearly and do one step at a time. This will reduce the chance of making a mistake.

Problem 5 : | *You Try*

Simplify the following.

a) $12\div3+1$

c) $8+\frac{1}{2}(10-4)$

b) $5-(3-11)^2$

d) $12\div(3+1)$

There are several types of grouping symbols that can be used besides parenthesis. One type is a fraction bar. If we have a fraction, the entire numerator and the entire denominator must be evaluated before we simpify the fraction. In these cases, we can simplify both the numerator and denominator at the same time.

| Problem 6 : | Worked Example |

Simplify: $\dfrac{2^4 - 3(-8)}{15 \div 5 - 1}$

Solution.

$$\frac{2^4 - 3(-8)}{15 \div 5 - 1} \qquad \text{Numerator:Apply exponent. Denominator: Divide}$$

$$= \frac{16 - 3(-8)}{3 - 1} \qquad \text{Numerator: Multiply. Denominator: Subtract}$$

$$= \frac{16 + 24}{2} \qquad \text{Numerator: Add}$$

$$= \frac{40}{2} \qquad \text{Simplify}$$

$$= 20 \qquad \text{Our Solution}$$

Note. Use extreme care when division involves zero. If zero is divided by a number other than zero, we get zero.

- Example: $0 \div 2 = 0$. To check, multiply the divisor by the quotient to get the dividend. Check: $0 \cdot 2 = 0$

- Example: $\dfrac{0}{-5} = 0$. Check: $0 \cdot -5 = 0$.

However, if a number, other than zero, is divided by 0, the result is undefined.

- Example: $2 \div 0$ is undefined. Check: there is no number such that 0 times that number will give us 2.

- Example: $\dfrac{-9}{0}$ is undefined. There is no number, such that if we multiply it by 0, we get -9.

Problem 7 : | *Worked Example*

Simplify: $\dfrac{3^2-(4+5)}{7+5}$

    Solution.

$\dfrac{3^2-(4+5)}{7+5}$ Numerator:Apply exponent. Denominator: Add

$=\dfrac{9-(4+5)}{12}$ Numerator: Perform operation inside parenthesis

$=\dfrac{9-9}{12}$ Numerator:Subtract

$=\dfrac{0}{12}$ Simplify

$=0$ Our Solution

Problem 8 : | *Worked Example*

Simplify: $\dfrac{(3)(-5)}{7+2-9}$

    Solution.

$\dfrac{(3)(-5)}{7+2-9}$ Numerator: multiply. Denominator: add

$=\dfrac{-15}{9-9}$ Denominator: Subtract

$=\dfrac{-15}{0}$ Cannot divide by 0

undefined   Our Solution.

Another type of grouping symbol is the absolute value. When we have an absolute value, we will evaluate everything inside the absolute value, as if it were a normal parenthesis. When the inside is completed, we will take the absolute value, or distance from zero. Note that

- $|5|=5$ because the distance from 0 to 5 is 5 units.

- $|-5|=5$ also because the distance from 0 to -5 is also 5 units.

Problem 9 : | *Worked Example*

Simplify: $3|4-8|+2|5+1|$

   Solution.

$$3|\underbrace{4-8}|+2|\underbrace{5+1}| \quad \text{Perform operation inside absolute value}$$

$$=3\,|\underbrace{-4}|+\,2\,|\underbrace{6}| \quad \text{Evaluate the absolute value}$$

$$=\underbrace{3(4)}+\underbrace{2(6)} \quad \text{Multiply}$$

$$=\underbrace{12\,+\,12} \quad \text{Add}$$

$$=24 \quad \text{Our Solution}$$

Problem 10 : | *Media/Class Example*

Simplify the following.

a) $\dfrac{(9-5)+(15-7)}{2(4+2)}$

c) $\dfrac{35-7(3+2)}{5^2-2^3}$

b) $|3-18|-|9+6|$

d) $-3|-2|$

Problem 11 :	*You Try*

Simplify the following.

a) $|4-9|-|-12+19|$

c) $|5^2-(3^2+4^2)|$

b) $\dfrac{-4^2-(4+2\cdot3)}{5+3(5-4)}$

d) $\dfrac{(4+5)(2-9)}{2^3-(2^2+4)}$

Practice Problems: *Order of Operations*

Simplify the following.

1. $-6 \cdot 4(-1)$

2. $\dfrac{1}{2} - \dfrac{2}{3} + \dfrac{3}{4}$

3. $(-6 \div 6)^2$

4. $7 - 4 \cdot \dfrac{4}{5}$

5. $9 - |2 - 10|$

6. $8 \div 4 \cdot 2^2$

7. $(3 - 7)^3$

8. $3 + (8) \div |4 - 2|$

9. $11 - (2 - 4)^3$

10. $6 - \dfrac{1}{2}(5 - 8)$

11. $5 + 3(2 - 6 \cdot 4)$

12. $5(-5 + 6) - 6^2$

13. $1 + 3|2 + 4|$

14. $\dfrac{-3 - 1}{-2 - (-2)}$

15. $\dfrac{1}{2} \cdot \dfrac{7}{3} - \left(\dfrac{4}{3}\right)^2$

16. $5^2 - (-8)^2$

17. $\dfrac{-10 - 6}{(-2)^2} - 5$

18. $4 - 2|3^2 - 16|$

19. $\dfrac{-5^2 + (-5)^2}{2^4 - 4 \cdot 3}$

20. $-6^2 + |-3 - 3|$

21. $\dfrac{2 - |7 + 2^2|}{4 \cdot 2 + 5 \cdot 3}$

22. $[-9 - (2 - 5)] \div (-6)$

23. $[-1 - (-5)] |3 + 2|$

24. $\dfrac{(-2 + 1) - (-3)}{-9 \cdot 2 - 3(-6)}$

 Rescue Roody!

25. Roody was asked to simplify $2 + 5(6 - 3)$. This is what he did but his work was marked incorrect. Help Roody understand what he did wrong.

$$\begin{aligned} 2 + 5(6 - 3) &= 2 + 5(3) \\ &= 7(3) \\ &= 21 \end{aligned}$$

26. Roody was asked to simplify $8 \div 4(2 - 7)$ but his answer does not match the answer in the back of the book. Help Roody.

$$\begin{aligned} 8 \div 4(2 - 7) &= 8 \div 4 - 5 \\ &= 2 - 5 \\ &= -3 \end{aligned}$$

1.2. ALGEBRAIC EXPRESSIONS

Objective: Evaluate algebraic expressions by substituting given values, and simplify algebraic expressions by distributing and combining like terms
Evaluating Algebraic Expressions

We are familiar with **numerical expressions:** numbers connected by some arithmetic operation. Examples are $3(5)$ or $-2(6-4^2)$

An **algebraic expression** is similar to a numerical expression, but it will also contain variables. What is a variable? A variable is a placeholder for an unknown value. Here are some examples: $3x$, $p-4.7$, $-\frac{2}{5}x+8y+25$.

A **term** is a number, a variable, or a product of numbers and variables. If there are two or more terms in an expression, they are separated by addition.

Consider the expression, $p-4.7$. This contains 2 terms, p and -4.7.
The expression $-\frac{2}{5}x+8y+25$ contains three terms, $-\frac{2}{5}x$, $8y$, and 25.

Looking at one term, such as $3x$, we say that 3 and x are **factors** of the term. 3 is also known as the **coefficient**.

Problem 1 :	*Class/Media Example*

What is the coeficient of the term $-\frac{2}{5}x$?

What is the coefficient of the term y?

What is the coefficient of the term $-\frac{p}{9}$?

World View Note: The term "Algebra" comes from the Arabic word al-jabr which means "reunion". It was first used in Iraq in 830 AD by Mohammad ibn-Musa al-Khwarizmi.

We can evaluate an expression by substituting a given value for each variable.

Problem 2 :	*Worked Example*

Evaluate the expression $p(q+6)$ when $p=3$ and $q=5$.

   Solution.

$$(3)((5)+6) \quad \text{Replace } p \text{ with } 3 \text{ and } q \text{ with } 5$$
$$=(3)((5)+6) \quad \text{Evaluate Parenthesis}$$
$$=(3)(11) \quad \text{Multiply}$$
$$=33 \quad \text{Our Solution}$$

Whenever a variable is replaced with a number, we will put the new number inside a set of parenthesis. Notice the 3 and 5 in the previous example are in parenthesis. This is to preserve operations that are sometimes lost in a simple replacement. Sometimes the parenthesis won't make a difference, but it is a good habit to always use them to prevent problems later.

Problem 3 :	*Class/Media Example*

a) Evaluate the expression $2.5p - q^2 + 4.8$ for $p = 10$ and $q = -6$

b) Evaluate the expression $\frac{x}{2} + 9xy$ for $x = -4$ and $y = \frac{1}{3}$

Problem 4 :	*You Try*

a) Evaluate the expression $-5\left(2n + \frac{3}{4}m\right)$ for $m = 12$ and $n = -7$

b) Evaluate the expression $\frac{a^3 - 1}{b^2 - 12}$ for $a = -3$ and $b = -4$

c) Evaluate the expression $\frac{xz}{4(3-z)}$ when $x = -6$ and $z = -2$

Simplifying Algebraic Expressions

We can simplify expressions by combining like terms. **Like terms** mean that the variables match exactly (exponents included). Examples of like terms are $3x$ and $-7x$, $3a^2$ and $8a^2$, 3 and -5. Examples of unlike terms are $2x, 3z$ and $3z^2$.

How does combining like terms work?

Consider shopping for fruit with a friend at a local farmer's market. You want to purchase 3 apples. Your friend wants to purchase 2 apples. How many total apples is this? 3 apples plus 2 apples give us 5 apples. Similarly, 3a plus 2a gives us 5a, or mathematically, $3a + 2a = 5a$. If we have like terms, add (or subtract) the coefficients, and keep the variables the same.

| **Problem 5 :** | *Worked Example* |

Simplify: $3x + 6x$

> **Solution.**

$$3x + 6x \quad \text{Add coefficients of } x$$
$$= 9x \quad \text{Our Solution}$$

What are unlike terms?

Returning to the shopping analogy, suppose you want to purchase 3 apples but your friend wants to purchase 2 loaves of bread. There is no way to describe apples and bread as the same thing. They are different items.

The same holds true with variables. We cannot combine $3a + 2b$ terms. These are unlike terms.

Why are x and x^2 not like terms?

Recall that raising to an exponent is another way of expressing repeated multiplication. For example,

$$3^2 = 3 \cdot 3 = 9 \quad \text{which is different from 3}$$
$$5^2 = 5 \cdot 5 = 25 \quad \text{which is different from 5}$$
$$x^2 = x \cdot x \quad \text{which is different from } x$$

Consequently, we cannot combine $x + x^2$. These are unlike terms.

| **Problem 6 :** | *Worked Example* |

Simplify: $2y + y^2$

> **Solution.**

Since there are no like terms, this expression cannot be simplified.

Problem 7 : | *Worked Example*

Simplify: $5x - 2y - 8x + 7y$

    Solution.

$$5x - 2y - 8x + 7y \quad \text{Combine like terms } 5x - 8x \text{ and } -2y + 7y$$
$$= -3x + 5y \quad \text{Our Solution}$$

Problem 8 : | *Worked Example*

Simplify: $8a - 3b + 7.1 - 2a + 2.5b - 3$

    Solution.

$$8a - 3b + 7.1 - 2a + 2.5b - 3 \quad \text{Combine like terms } 8a - 2a, -3b + 2.5b \text{ and } 7.1 - 3$$
$$= 6a - 0.5b + 4.1 \quad \text{Our Solution}$$

Problem 9 : | *Class/Media Example*

Simplify: $7s - 4p + 12s - 18p$

Problem 10 : | *You Try*

Simplify.

a) $\dfrac{1}{2}x + 3y - 4 - 5x + \dfrac{8}{3}$

b) $2n + 2n^2$

The distributive property is another tool we use to simplify algebraic expressions.

$$\text{Distributive Property: } a(b+c) = ab + ac$$

Problem 11 : | *Worked Example*

Multiply: $4\left(2x - \frac{9}{2}\right)$

Solution.

$$4\left(2x - \frac{9}{2}\right) \qquad \text{Apply the distributive property}$$
$$= 4(2x) - 4\left(\frac{9}{2}\right) \qquad \text{Multiply each term by } 4$$
$$= 8x - 18 \qquad \text{Our Solution}$$

Problem 12 : | *Worked Example*

Multiply: $-7(5x - 6)$

Solution.

$$-7(5x - 6) \qquad \text{Apply the distributive property}$$
$$= -7(5x) + (-7)(-6) \qquad \text{Multiply each term by } -7$$
$$= -35x + 42 \qquad \text{Our Solution}$$

Note. A common error in distributing is a sign error. Be very careful with the signs!

Problem 13 : | *Class/Media Example*

Multiply: $0.2(10q - 4)$

When a negative sign is in front of the parenthesis, distribute -1.

Problem 14 : | *Worked Example*

Simplify: $-(4x - 5y + 6)$

Solution.

$$-(4x - 5y + 6) \qquad \text{Apply distributive property}$$
$$= -1(4x) + (-1)(-5y) + (-1)(6) \qquad \text{Multiply each term by } -1$$
$$= -4x + 5y - 6 \qquad \text{Our Solution}$$

Note. Distributing the negative in front of the parenthesis has the same effect as changing the sign of each term inside the parenthesis.

Problem 15 : | *You Try*

a) Simplify: $-(3x+4y-2)$ b) Simplify: $4\left(-5k+\dfrac{1}{3}p\right)$

The next examples require both distribution and combining like terms.

Problem 16 : | *Worked Example*

Simplify: $5+3(2x-4)$

   Solution.

$$5+3(2x-4) \quad \text{Distribute } 3$$
$$=5+6x-12 \quad \text{Combine like terms } 5,-12$$
$$=6x-7 \quad \text{Our Solution}$$

Problem 17 : | *Worked Example*

Simplify: $3x-2(4x-5)$

   Solution.

$$3x-2(4x-5) \quad \text{Distribute } -2$$
$$=3x-8x+10 \quad \text{Combine like terms } 3x,-8x$$
$$=-5x+10 \quad \text{Our Solution}$$

Problem 18 : | *Worked Example*

Simplify: $4(3x-8)-(2x-7)$

   Solution.

$$4(3x-8)-(2x-7)$$
$$=4(3x-8)-1(2x-7) \quad \text{Distribute 4 into first parenthesis, -1 into second}$$
$$=12x-32-2x+7 \quad \text{Combine like terms } 12x,-2x \text{ and } -32,7$$
$$=10x-25 \quad \text{Our Solution}$$

Problem 19 : | *Class/Media Example*

Simplify: $10x - 16 - 6(4k + 3)$

Problem 20 : | *You Try*

Simplify the following:

a) $2(3p - 4) + 5$

d) $4 - \dfrac{1}{2}(8k - 1)$

b) $(7m - 3) + (5m + 6)$

e) $(3g - 5) - (7 - 8g)$

c) $6(9a + 5) + 4(b + 3)$

f) $\dfrac{2}{3}(9n + 6) - \dfrac{1}{4}(12 - 8n)$

Commutative and Associative Properties

There are two properties that help us simplify expressions.

Commutative Property of Addition (and Multiplication) - order of numbers can be changed without changing the result, that is,

$$a + b = b + a \quad and \quad ab = ba$$

Associative Property of Addition (and Multiplication) - numbers can be grouped in any way without changing the result, that is,

$$a + (b + c) = (a + b) + c \quad and \quad a(bc) = (ab)c$$

Problem 21 : | *Worked Example*

Simplify: $(3x-1)4$

> Solution.

$$(3x-1)4 \quad \text{Reorder using commutative property}$$
$$=4(3x-1) \quad \text{Distribute 4}$$
$$=12x-4 \quad \text{Our Solution}$$

Problem 22 : | *Worked Example*

Simplify: $4\left(\dfrac{3}{2}(p-7)\right)$

> Solution.

$$4\left(\frac{3}{2}(p-7)\right) \quad \text{Regroup product using associative property}$$

$$=\underbrace{\left(4\cdot\frac{3}{2}\right)}(p-7) \quad \text{Multiply 4 and } \frac{3}{2}$$
$$=6(p-7) \quad \text{Distribute 6}$$

$$=6p-42 \quad \text{Our Solution}$$

Problem 23 : | *Class/Media Examples*

Simplify the following.

a) $9\left(\dfrac{x+1}{3}\right)$

b) $-\dfrac{5}{3}(6(y+4))$

Problem 24 : | *You Try*

Simplify the following.

a) $-12\left(\dfrac{2k+3}{4}\right)-5\left(\dfrac{3k+7}{5}\right)$

b) $\dfrac{3}{4}\left(-8\left(\dfrac{1}{2}x+5\right)\right)$

Problem 2 : | *Media/Class Example*

Solve $5 = 8 + c$ for c. Be sure to check your answer.

Problem 3 : | *You Try*

Solve the following equations. Be sure to check your answer.

a) $4 + g = {}^-8$ b) $7 = k + 9$

The same process is used in the following examples. Notice that this time we are getting rid of a negative number by adding.

Problem 4 : | *Worked Example*

Solve $x - 5 = 4$ for x.

   Solution.

$$x - 5 = 4 \quad \text{Isolate the variable, } x$$
$$\underline{+5 \, +5} \quad \text{Add 5 to both sides of equation}$$
$$x = 9 \quad \text{Our Solution!}$$

Check if $x = 9$ is the correct solution. Substitute $x = 9$ into the equation.

$$(9) - 5 \overset{?}{=} 4$$
$$4 = 4\checkmark$$

This verifies that $x = 9$ is the solution.

Problem 5 : | *Media/Class Example*

Solve the following equations. Be sure to check your answer.

a) $-10 = w - 7$ b) $9 = {}^-3 + y$

Problem 6 : | *You Try*

Solve the following equations. Be sure to check your answer.

a) $-4+x=-7$ b) $p-\frac{1}{2}=3$ c) $1.5=n-2.3$

Multiplication Property of Equality
Given an equation $A=B$, then $AC=BC$.

This means that we can multiply both sides of an equation by the same factor and both sides will remain equal. Consequently, the solution is unchanged.

Division Property of Equality
Given an equation $A=B$, then $\dfrac{A}{C}=\dfrac{B}{C}$, for any $C \neq 0$.

This means that we can divide both sides of an equation by anything other than 0 and both sides will remain equal. Consequently, the solution is unchanged.

With a multiplication problem, we can isolate the variable by dividing both sides of the equation by the variable's coefficient. Consider the following example.

Problem 7 : | *Worked Example*

Solve $4m=20$ for m.

   Solution.

$$4m=20 \quad \text{Isolate the variable } m$$

$$\frac{4m}{4}=\frac{20}{5} \quad \text{Divide both sides by 4, the coefficient of } m$$

$$m=5 \quad \text{Our solution!}$$

Check if $m=5$ is the correct solution. Substitute $m=5$ into the equation.

$$4(5) \overset{?}{=} 20$$
$$20 = 20\checkmark$$

This verifies that $m=5$ is the solution.

With multiplication problems, it is very important that care is taken with signs. If x is multiplied by a negative number, then we will divide by a negative negative as shown in the following example.

Problem 8 : | Worked Example

Solve $-5x = 30$ for x.

> **Solution.**

$$-5x = 30 \quad \text{Isolate the variable } x$$

$$\frac{-5x}{-5} = \frac{30}{-5} \quad \text{Divide both sides by } -5, \text{ the coefficient of } x$$

$$x = -6 \quad \text{Our Solution!}$$

Check if $x = -6$ is the correct solution. Substitute $x = -6$ into the equation.

$$-5(-6) \stackrel{?}{=} 30$$
$$30 = 30 \checkmark$$

This verifies that $x = -6$ is the solution.

Problem 9 : | Worked Example

Solve for $-x = 8$ for x.

> **Solution.**

Note that we are solving for x. There is a negative sign in front of the x which needs to be cleared so that the variable x is isolated.

Method #1: Multiply each side by -1:

$$(-1)(x) = (-1)(8)$$
$$x = -8$$

Method #2: Divide each side by -1:

$$\frac{-x}{-1} = \frac{8}{-1}$$

$$x = -8$$

Both methods yield the same answer: $x = -8$.
Check is $x = -8$ is the correct solution. Substitute $x = -8$ into the equation.

$$-(-8) = 8 \checkmark$$

This verifies that $x = -8$ is the solution.

Problem 10 : | *Class/Media Example*

Solve the following equations. Be sure to check your answer.

a) $-x = -24$　　　　　　b) $-6p = 20$　　　　　　c) $4.2 = 7w$

When the coefficient is a fraction, we multiply by the reciprocal to isolate the variable.

Problem 11 : | *Worked Example*

Solve $\dfrac{x}{5} = -3$ for x.

　　Solution.

$$\frac{x}{5} = -3 \qquad \text{Coefficient of } x \text{ is } \frac{1}{5}$$

$$(5)\frac{x}{5} = -3(5) \qquad \text{Multiply both sides by } 5, \text{ the reciprocal of } \frac{1}{5}$$

$$x = -15 \qquad \text{Our Solution!}$$

Check if $x = -15$ is the correct solution. Substitute $x = -15$ into the equation.

$$\frac{-15}{5} \stackrel{?}{=} -3$$
$$-3 = -3\checkmark$$

This verifies that $x = -15$ is the solution.

Problem 12 : | *Class/Media Example*

Solve the following. Be sure to check your answer.

a) $\dfrac{m}{-7} = -2$　　　　　　　　　　　b) $\dfrac{3}{5}p = 9$

Problem 6 : | *You try*

Solve the following equations. Be sure to check your solutions.

a) $5x+1=12-x$ b) $4x+2=10x+1$

Problem 7 : | *Worked example*

Solve $3(x-2)=4(x+1)$

$$3(x-2) = 4(x+1)$$
$$3x-6 = 4x+4 \qquad \text{Distribute}$$

$$3x-6 = 4x+4$$
$$\underline{-4x \qquad -4x} \qquad \text{Subtract } 4x \text{ from each side}$$
$$-x-6 = 4$$

$$-x-6 = \quad 4$$
$$\underline{+6 \qquad +6} \qquad \text{Add 6 to each side}$$
$$-x = 10$$

$$(-1)(-x) = (-1)10 \qquad \text{Multiply each side by } (-1)$$
$$x = -10 \qquad \text{Our solution}$$

Let's check our solution:

$$3(-10-2) \stackrel{?}{=} 4(-10+1)$$

$$3(-12) \stackrel{?}{=} 4(-9)$$
$$-36 = -36 \qquad \checkmark$$

Note. In our last example we have not isolated completely the variable when we reached the point where

$$-x=10$$

Our goal is to solve this equation for x, not $-x$, that is why we need to multiply each side by (-1).

Problem 8 : | *Class/media example*

Solve $5(3-r)+2=3(r+2)$. Be sure to check your solution.

Problem 9 : | *Class/media example*

Solve $3(2p+8)=p+24$. Be sure to check your solution.

Problem 10 : | *You try*

Solve the followng equations. Be sure to check your solutions.

   a) $4(x+3)-2(x-1)=0$ b) $5(x+2)=3(x-3)$

Problem 11 : | *Worked Example*

Solve $3(x-2)-x=2x+1$

$$3(x-2)-x = 2x+1$$
$$3x-6-x = 2x+1 \qquad \text{perform the multiplication}$$
$$2x-6 = 2x+1 \qquad \text{combine like terms}$$

$$
\begin{array}{rcl}
2x-6 &=& 2x+1 \\
-2x & & -2x \\
\hline
-6 &=& 1 \qquad \text{!!!!!}
\end{array}
$$

$$\text{subtract } 2x \text{ from each side}$$

The last statement, $-6=1$, is clearly false. We can interpret this as follows:

- when solving an equation we are looking for the value(s) of our variable that make the equality true.

- if we reach a false statement, such as $-6 = 1$, this says that **there is no value of x that makes the equality true**

- this means that our equation has **No Solution**.

Problem 12 : | *Worked Example*

Solve $4(x+3)=2x+2(x+6)$

$$4(x+3) = 2x+2(x+6)$$
$$4x+12 = 2x+2x+12 \qquad \text{perform the multipliations on each side}$$
$$4x+12 = 4x+12 \qquad \text{combine like terms}$$

Something interesting happened in this example as well:

- We can see that our last line is an equality that will be true **for any value of x.**

- This means that any real number is a solution for this equation.

- We will express this by saying that the solution consists of **all real numbers**, R

Notice that if we continue working in our last example, trying to isolate the variable, we obtain

$$
\begin{array}{rcl}
4x+12 &=& 4x+12 \\
-4x & & -4x \\
\hline
12 &=& 12
\end{array}
$$

$$\text{subtract } 4x \text{ from each side}$$

This last statement, $12=12$, is true, regardless of the value of x. This also tells us that the solution to our original equation is the set of all real numbers, R.

When solving an equation

- if we obtain a false statement, such as $-6=11$, the equation has **No Solution**.

- if we obtain a true statement, regardless of the value of our variable, such as in $12=12$, then the solution for the equation consists of **all real numbers**, R.

Problem 13 : | *Class/media example*

Solve the following equations. Be sure to check your solutions, if applicable.

 a) $3(x-4)+2= 2(x-3)+x-4$ b) $4(2-x)+7x =3(x+4)$

Problem 14 : | *You try*

Solve the following equations. Be sure to check your solutions, if applicable.

 a) $5(m+2)+1= 2(m-3)+3m$ b) $4(1-y)+4(y-1) =0$

Problem 15 : | *Worked example*

Solve $\frac{1}{2}x+3=\frac{3}{4}$

Method 1: Work directly with the fractions

$$\frac{1}{2}x+3 \ = \ \frac{3}{4}$$

$$\begin{array}{rcl} \frac{1}{2}x+3 & = & \frac{3}{4} \\ \underline{\quad -3} & & \underline{-3} \end{array} \qquad \text{subtract 3 from each side}$$

$$\frac{1}{2}x \ = \ \frac{3}{4}-3$$

$$\frac{1}{2}x \ = \ \frac{3-12}{4} \qquad \text{add using the LCD on the right side}$$

$$\frac{1}{2}x \ = \ \frac{-9}{4}$$

$$2\left(\frac{1}{2}x\right) \ = \ 2\left(\frac{-9}{4}\right) \qquad \text{multiply by 2 each side}$$

$$x \ = \ \frac{-9}{2} \qquad \text{our solution}$$

Method 2: Multiply each side by the LCD of all fractions appearing in the equation.

$$\frac{1}{2}x+3 \quad = \quad \frac{3}{4}$$

$$4\left(\frac{1}{2}x+3\right) \ = \ 4\left(\frac{3}{4}\right) \qquad \text{multiply each side bt the LCD (4)}$$

$$2x+12 \ = \ 3$$

$$\begin{array}{rcl} 2x+12 & = & 3 \\ \underline{\quad -12} & & \underline{-12} \\ 2x & = & -9 \end{array} \qquad \text{subtract 12 from each side}$$

$$\frac{2x}{2} \ = \ \frac{-9}{2} \qquad \text{divide each side by 2}$$

$$x \ = \ \frac{-9}{2} \qquad \text{our solution.}$$

Let's check our solution:

$$\frac{1}{2}\left(\frac{-9}{2}\right) + 3 \overset{?}{=} \frac{3}{4} \quad \text{substitute}$$

$$\frac{-9}{4} + 3 \overset{?}{=} \frac{3}{4} \quad \text{multiply}$$

$$\frac{-9+12}{4} \overset{?}{=} \frac{3}{4} \quad \text{add using the LCD}$$

$$\frac{3}{4} = \frac{3}{4} \quad \checkmark$$

Problem 16 : | *Worked example*

Solve $\frac{3}{4}(x-2) = x + \frac{5}{2}$

Method 1: Work directly with the fractions

$$\frac{3}{4}(x-2) = x + \frac{5}{2}$$

$$\frac{3x}{4} - \frac{6}{4} = x + \frac{5}{2} \quad \text{distribute}$$

$$\frac{3x}{4} = x + \frac{5}{2} + \frac{3}{2} \quad \text{add } \frac{3}{2} \text{ to each side}$$

$$\frac{3x}{4} = x + \frac{8}{2} \quad \text{combine terms}$$

$$\begin{array}{r} \frac{3x}{4} = x + \frac{8}{2} \\ \underline{-x \qquad -x} \\ \frac{3x}{4} - x = 4 \end{array} \quad \text{subtract } x \text{ to each side}$$

$$\frac{3x - 4x}{4} = 4 \quad \text{add using the LCD}$$

$$\frac{-x}{4} = 4 \quad \text{combine terms}$$

$$(-4)\left(\frac{-x}{4}\right) = (-4)4 \quad \text{multiply each side by } (-4)$$

$$x = -16 \quad \text{Our solution}$$

Method 2: Multiply each side by the LCD of all fractions appearing in the equation.

$$\frac{3}{4}(x-2) = x+\frac{5}{2}$$

$$4\left(\frac{3}{4}(x-2)\right) = 4\left(x+\frac{5}{2}\right) \quad \text{multiply each side by the LCD}$$

$$3(x-2) = 4\left(x+\frac{5}{2}\right) \quad \text{simplify the fractions}$$

$$3x-6 = 4x+\frac{20}{2} \quad \text{distribute the multiplication}$$

$$3x-6 = 4x+10 \quad \text{simplify the fractions}$$

$$\begin{array}{rcl} 3x-6 &=& 4x+10 \\ \underline{+6} & & \underline{+6} \\ 3x &=& 4x+16 \end{array} \quad \text{add 6 to each side}$$

$$\begin{array}{rcl} 3x &=& 4x+16 \\ \underline{-4x} & & \underline{-4x} \\ -x &=& 16 \end{array} \quad \text{subtract } 4x \text{ from each side}$$

$$(-1)(-x) = (-1)16 \quad \text{multiply both sides by -1}$$

$$x = -16 \quad \text{our solution.}$$

Let's check our solution:

$$\frac{3}{4}(-16-2) = -16+\frac{5}{2} \quad \text{substitute}$$

$$\frac{3}{4}(-18) \overset{?}{=} \frac{-32+5}{2} \quad \text{combine terms and add using the LCD}$$

$$\frac{-54}{4} \overset{?}{=} \frac{-27}{2} \quad \text{perform the operations}$$

$$\frac{-27}{2} = \frac{-27}{2} \quad \checkmark$$

Problem 17 : | *Class/Media Example*

Solve the equation $\frac{3}{5}(x-3)+\frac{1}{2}=4x+\frac{1}{4}$

Problem 18 : | *You try*

Solve the equation $\frac{3}{5}(x-2)=\frac{1}{2}$

Problem 19 : | *You try*

Solve the equation $\frac{3}{2}(x-2)=\frac{x-2}{3}+1$

World View Note: Persian mathematician Omar Khayyam would solve algebraic problems geometrically by intersecting graphs rather than solving them algebraically.

Practice Problems: *General Linear Equations*

Solve each equation.

1. $5 + \dfrac{n}{4} = 4$

2. $102 = -7r + 4$

3. $-8n + 3 = -77$

4. $0 = -6v$

5. $-8 = \dfrac{x}{5} - 6$

6. $0 = -7 + \dfrac{k}{2}$

7. $-12 + 3x = 0$

8. $24 = 2n - 8$

9. $2 = -12 + 2r$

10. $\dfrac{b}{3} + 7 = 10$

11. $\dfrac{3}{5}(1 + p) = \dfrac{21}{20}$

12. $0 = -\dfrac{5}{4}\left(x - \dfrac{6}{5}\right)$

13. $\dfrac{3}{4} - \dfrac{5}{4}m = \dfrac{113}{24}$

14. $\dfrac{635}{72} = -\dfrac{5}{2}\left(-\dfrac{11}{4} + x\right)$

15. $2b + \dfrac{9}{5} = -\dfrac{11}{5}$

16. $\dfrac{3}{2}\left(\dfrac{7}{3}n + 1\right) = \dfrac{3}{2}$

17. $-a - \dfrac{5}{4}\left(-\dfrac{8}{3}a + 1\right) = -\dfrac{19}{4}$

18. $\dfrac{55}{6} = -\dfrac{5}{2}\left(\dfrac{3}{2}p - \dfrac{5}{3}\right)$

19. $\dfrac{16}{9} = -\dfrac{4}{3}\left(-\dfrac{4}{3}n - \dfrac{4}{3}\right)$

20. $-\dfrac{5}{8} = \dfrac{5}{4}\left(r - \dfrac{3}{2}\right)$

21. $-2(8n - 4) = 8(1 - n)$

22. $-3(-7v + 3) + 8v = 5v - 4(1 - 6v)$

23. $-7(x - 2) = -4 - 6(x - 1)$

24. $-6(8k + 4) = -8(6k + 3) - 2$

25. $-2(1 - 7p) = 8(p - 7)$

26. $-4(1 + a) = 2a - 8(5 + 3a)$

27. $-6(x - 3) + 5 = -2 - 5(x - 5)$

28. $-(n + 8) + n = -8n + 2(4n - 4)$

29. $-5(x + 7) = 4(-8x - 2)$

30. $8(-8n + 4) = 4(-7n + 8)$

MID-CHAPTER CHECK-UP

Simplify the following expressions.

1. $-2(6-9)^2-5$

2. $\dfrac{-12-6^2}{-15+5}$

3. $3(4k-1)-2(6k+4)$

4. $5-\dfrac{3}{4}(12n-8)$

Evaluate each expression using the values given.

5. $a^2-(b-4)^2$; use $a=-3, b=-6$

6. $\dfrac{8+2m}{n^2+1}$; use $m=-4, n=2$

Solve the following equations.

7. $m+7=-2$

8. $\dfrac{m}{3}=9$

9. $\dfrac{2}{5}m+10=0$

10. $2x-8=6x+12$

11. $2(9-y)=y-3(y-6)$

12. $\dfrac{1}{2}(6-m)=\dfrac{1}{3}(2m+9)$

1.5. Linear Inequalities

Objective: Solve, graph, and give interval notation for the solution to linear inequalities.

When we have an equation such as $x = 4$, there is a specific value assigned to the variable, x. With inequalities we will give a range of values for our variable. To do this we will use one of the following symbols:

>   Greater than

$\geq$ Greater than or equal to

<   Less than

$\leq$ Less than or equal to

An inequality such as $x < 4$ means that our variable can be any number smaller than 4 such as $-2, 0, 3, 3.9$ or even 3.999999999 while an inequality such as $x \geq -2$ means that our variable can be any number greater than or equal to -2 such as $5, 0, -1, -1.99999$, including -2.

Because we don't have one set value for our variable, it is often useful to draw a picture of the solutions on a number line. Once the graph is drawn, we will also write our solution in interval notation. Interval notation gives two numbers, the first is the smallest value, the second is the largest value. If there is no largest value, we will use ∞ (infinity). If there is no smallest value, we will use $-\infty$ (negative infinity). Inequalities that do not have an equal sign will use open circles for graphs and parenthesis, (or) for interval notation. Inequalities that involve the equal sign will use closed circles for graphs and brackets, [or] for interval notation. The number line will be shaded in the appropriate region to denote all possible solutions of the variable.

Note. Infinity is not a number. It is an idea of something without end. When writing your solution in interval notation, infinity will always be enclosed by a parenthesis and not a bracket.

World View Note: English mathematician Thomas Harriot first used the above symbols in 1631. However, they were not immediately accepted as symbols such as $=$ and $\exists$ were already coined by another English mathematician, William Oughtred.

Problem 1 : | Worked Example

Graph $x < 2$ on the number line and write your solution in interval notation.

> Solution.

We want all solutions of x that are smaller than 2, but not including 2. We will use an open circle on 2, to show that 2 is not included and shade everything to the left of 2 to show all solutions less than 2.

Interval notation is $(-\infty, 2)$ since there is no smallest value and the largest value is 2.

Problem 2 : | Worked Example

Graph $y \geq -1$ on the number line and write your solution in interval notation.

> Solution.

We want all solutions of y that are greater than -1, and including -1. We will use a closed circle around -1, to show that -1 is included and shade everything to the right of -1 to show all solutions greater than -1.

Interval notation is $[-1, \infty)$ since the smallest value is -1 and there is no largest value.

Problem 3 : | Meida/Class Example

a) Graph $p \leq 5$ on the number line and write your solution in interval notation.

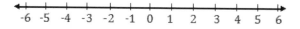

b) Graph $-3 > z$ on the number line and write your solution in interval notation.

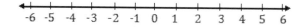

It is important to be careful when the inequality is written backwards as in the previous example. Students often draw their graphs the wrong way when this is the case. One way to avoid graphing the inequality the wrong way is to rewrite $-3 > z$ as $z < -3$.

Problem 4 : | *You Try*

a) Graph $-4 < c$ on the number line and write your solution in interval notation.

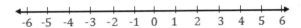

b) Graph $n \le 0$ on the number line and write your solution in interval notation.

Problem 5 : | *Worked Example*

Find the inequality that represents the graph below and write the solution in interval notation.

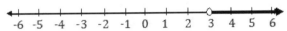

   Solution.

A parenthesis is used, that means 3 is not included in the solution. Shaded portion is to the right of 3. Solution must be $x > 3$. Interval notation: $(3, \infty)$.

Problem 6 : | *Worked Example*

Find the inequality that represents the graph below and write the solution in interval notation.

   Solution.

A bracket is used, that means -4 is included in the solution. Shaded portion is to the left of -4. Solution must be $x \le -4$. Interval notation: $(-\infty, -4]$.

Problem 7 : | *You Try*

Find the inequality that represents each graph below and write the solution in interval notation.

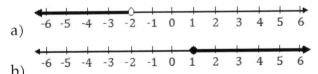

a)

b)

Solving Linear Inequalities

Solving inequalities is very similar to solving equations. Consider the following inequalities. Notice what happens to the inequality sign as we add, subtract, multiply and divide by both positive and negative numbers to keep the statement a true statement.

Start with	Operation Performed	Result	Inequality Sign
6>2	Add 2 to each side	8>4	Correct
6>2	Subtract 5 from each side	1>−3	Correct
6>2	Add −4 to each side	2>−2	Correct
6>2	Subtract −8 from each side	14>10	Correct
6>2	Multiply each side by 7	42>14	Correct
6>2	Divide each side by 6	$1 > \frac{1}{3}$	Correct
6>2	Multiply each side by −3	−18>−6	*Incorrect*
6>2	Divide each side by −1	−6>−2	*Incorrect*

As illustrated above, the inequality is preserved when we add or subtract to each side of the inequality. The same is true when we multiply or divide by a positive number. But if we multiply or divide by a negative number, the inequality reverses directions.

Problem 8 : | Worked Example

Solve the inequality: $y-3<-5$. Graph on a number line and write your solution in interval notation.

Solution.

$$
\begin{array}{ll}
y-3 \; < \; -5 & \\
\underline{+3 \quad\;\; +3} & \text{Add 3 to each side of inequality} \\
y \; < \; -2 & \text{Our solution}
\end{array}
$$

$(-\infty,-2)$ Interval Notation

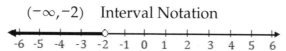

| Problem 9 : | *Worked Example* |

Solve the inequality: $-5m \geq 20$. Graph on a number line and write your solution in interval notation.

Solution.

$$-5m \geq 20 \quad \text{Divide each side by } -5; \text{Reverse inequality symbol}$$

$$\frac{-5m}{-5} \leq \frac{20}{-5}$$

$$m \leq -4 \quad \text{Our solution}$$

$$(-\infty, -4] \quad \text{Interval Notation}$$

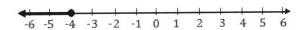

| Problem 10 : | *Media/Class Example* |

Solve the following inequality. Graph on a number line and write your solution in interval notation.

a) $3 \geq m + 9$

b) $4 < \dfrac{n}{6}$

| Problem 11 : | *You Try* |

Solve the following inequality. Graph on a number line and write your solution in interval notation.

a) $\dfrac{2}{3} + g \leq \dfrac{1}{5}$

b) $5 > -\dfrac{1}{2}k$

Problem 12 : | *Worked Example*

Solve the inequality: $5-2x \geq 11$. Graph on a number line and write your solution in interval notation.

Solution.

$$5-2x \geq 11$$

$$\underline{-5 \qquad -5} \quad \text{Subtract 5 from each side}$$

$$-2x \geq 6$$

$$\frac{-2x}{-2} \leq \frac{6}{-2} \quad \text{Divide each side by -2; Reverse the Inequality symbol}$$

$$x \leq -3 \quad \text{Our solution}$$

$$(-\infty, -3] \quad \text{Interval Notation}$$

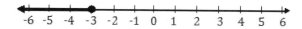

Problem 13 : | *Worked Example*

Solve the inequality: $11x-12 < 4(3x-7)+8$. Graph on a number line and write your solution in interval notation.

Solution.

$$11x-12 < 4(3x-6)+8 \quad \text{Distribute the 4}$$

$$11x-12 < 12x-24+8 \quad \text{Combine like terms}$$

$$11x-12 < 12x-16$$

$$11x-12 < 12x-16 \quad \text{Subtract } 12x \text{ from each side}$$

$$\underline{-12x \qquad -12x}$$

$$-x-12 < -16$$

$$-x-12 < -16 \quad \text{Add 12 to each side}$$

$$\underline{+12 \quad +12}$$

$$-x < -4$$

$$\frac{-x}{-1} > \frac{-4}{-1} \quad \text{Divide each side by -1; Reverse the inequality}$$

$$x > 4 \quad \text{Our solution}$$

$$(4, \infty) \quad \text{Interval Notation}$$

Problem 14 : | *Media/Class Example*

Solve the following inequality. Graph on a number line and write your solution in interval notation.

a) $3(y-2) \geq -(y+6)$

b) $6.8p + 4.2 < 3.5p - 12.3$

Problem 15 : | *You Try*

Solve the following inequality. Graph on a number line and write your solution in interval notation.

a) $4 - (g+3) < 5g$

b) $\dfrac{4}{5}w - \dfrac{1}{2} \geq \dfrac{3}{10}w + 2$

Practice Problems: *Linear Inequalities*

Complete the table.

	Inequality Notation	Number Line Graph	Interval Notation
1.	$n > -5$		
2.			$[4, \infty)$
3.		number line: open circle at -2, shaded to left	
4.	$1 \geq k$		
5.			$(-\infty, -7)$
6.		number line: closed circle at 1, shaded to left	
7.	$-6 \leq p$		
8.		number line: open circle at -2, shaded to right	
9.			$(-\infty, 8]$
10.	$x < 4$		

Solve each inequality, graph on a number line and write the solution in interval notation.

11. $2 + r < 3$

12. $-7n - 10 \geq 60$

13. $-8(n - 5) \geq 0$

14. $8 + \dfrac{n}{3} > 6$

15. $\dfrac{6 + x}{12} \leq -1$

16. $24 \geq -6(m - 6)$

17. $-2(3 + k) < -44$

18. $-r - 5(r - 6) < -18$

19. $11 \geq 8 + \dfrac{x}{2}$

20. $24 + 4b < 4(1 + 6b)$

21. $-8(2 - 2n) \geq -16 + n$

22. $4 + 2(a + 5) < -2(-a - 4)$

23. $-(k - 2) > -k - 20$

24. $-36 + 6x > -8(x + 2) + 4x$

25. $-5v - 5 \leq -5(4v + 1)$

Rescue Roody!

26. Roody was told to solve the inequality $6 - 2(m + 3) \geq 7 - 2m$ but he keeps getting stuck. Help Roody. Here is his work.

$$6 - 2(m+3) \geq 7 - 2m$$
$$6 - 2m - 6 \geq 7 - 2m$$
$$-2m \geq 7 - 2m$$
$$\underline{+2m \qquad + 2m}$$
$$0m \geq 7$$

1.6. ABSOLUTE VALUE EQUATIONS

Objective: Solve linear absolute value equations.

Recall that the absolute value of a number is its distance from 0 on the number line. For example, $|3| = 3$ and $|-3| = 3$ since the numbers 3 and -3 are 3 units from 0 on the number line.

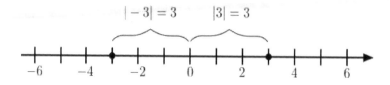

When dealing with equations involving absolute value, there can be more than one solution because the inside of the absolute value could be positive or negative. We must account for both possibilities.

Problem 1 : | *Worked Example*

Solve: $|x| = 3$.

> **Solution.**
>
> $$|x| = 3 \qquad x \text{ is 3 units from 0 on the number line}$$
> $$x = -3 \text{ or } x = 3 \qquad \text{Our Solution}$$

Notice that we have considered two possibilities as illustrated on the number line below.

Problem 2 : | *Media/Class Example*

Solve each of the following equation. Be sure to check your answer.

a) $|x| = 7$ b) $|w| = 1.5$

Problem 3 : | You Try

Solve each of the following equation. Be sure to check your answer.

a) $|x|=9$ b) $|y|=\frac{5}{2}$ c) $|n|=0$

Often we will have a more complex expression within the absolute value symbol. We still consider the distance the expression is from 0 on the number line. This is illustrated below on the number line.

Solve: $|\square|=a$

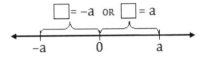

Problem 4 : | Worked Example

Solve each of the following equations. Be sure to check your answers.

a) $|3x|=12$ b) $|2x-1|=7$

Solution.

a) The expression $3x$ is 12 units from 0 on the number line. This results in the following two linear equations:

$$3x=-12 \quad \text{or} \quad 3x=12$$
$$x=-4 \quad \text{or} \quad x=4$$

Verify the solutions.

$$\text{Substitute } x=-4 \qquad \text{Substitute } x=4$$
$$|3(-4)|\overset{?}{=}12 \qquad |3(4)|\overset{?}{=}12$$
$$|-12|=12\checkmark \qquad |12|=12\checkmark$$

b) The expression $2x-1$ is 7 units from 0 on the number line. This results in the following two linear equations:

$$2x-1=-7 \quad \text{or} \quad 2x-1=7$$
$$2x=-6 \quad \text{or} \quad 2x=8$$
$$x=-3 \quad \text{or} \quad x=4$$

Verify the solutions.

$$\text{Substitute } x = -3$$

$$|2(-3) - 1| \overset{?}{=} 7$$

$$|-6 - 1| \overset{?}{=} 7$$

$$|-7| = 7 \checkmark$$

$$\text{Substitute } x = 4$$

$$|2(4) - 1| \overset{?}{=} 7$$

$$|8 - 1| \overset{?}{=} 7$$

$$|7| = 7 \checkmark$$

Problem 5 : | *Media/Class Example*

Solve each equation. Be sure to check your answer.

a) $|7x| = 21$

b) $|2y - 1| = 3$

Problem 6 : | *You Try*

Solve each equation. Be sure to check your answer.

a) $|v - 3| = 3$

c) $|4t - 3| = 11$

b) $|-x| = 9$

d) $|1 - p| = 0$

Remember the value of the absolute value must always be non-negative. Notice what happens in the next example.

Problem 7 : | *Worked Example*

Solve: $|x|=-2$.

   Solution.

Since the absolute value of any number cannot be negative, this equation does not have a solution! Our answer, therefore, is NO solution.

Note. If you think the solution is $x=-2$ or $x=2$, check your answer by substituting the value of x into the equation above. Neither solution is going to work.

$$\text{Substitute } x=-2 \qquad \text{Substitute } x=2$$
$$|-2| \neq -2 \qquad\qquad |2| \neq -2$$

Problem 8 : | *Media/Class Example*

Solve the following equation: $|1+3x|=-1$

Problem 9 : | *You Try*

Solve each of the following equations.

a) $|1-y|=-8$ 　　　　　b) $|2x-6|=-1$ 　　　　　c) $|x-5|=2$

Practice Problems: *Absolute Value Equations*

Solve each equation, if possible. Check your answer, if appropriate.

1. $|x| = 3$

2. $|y| = 0$

3. $|k| = -1$

4. $|5p| = 25$

5. $|9n + 8| = 46$

6. $|6 - 2c| = 24$

7. $|-3w| = 6$

8. $\left|\dfrac{4b + 10}{8}\right| = 3$

9. $|8(x + 7)| = 8$

10. $|y + 1| = 8$

11. $\left|\dfrac{3m + 1}{2}\right| = 5$

12. $\left|\dfrac{2}{3}x\right| = \dfrac{38}{9}$

13. $|4 + x| = -9$

14. $|3 - 4t| = 0$

15. $|-2g| = 10$

16. $|8 - k| = 8$

17. $|4c - 3| = 7$

18. $|m + 5| = -8$

1.7. Formulas

Objective: Solve formulas for a given variable.

A formula establishes a relationship between two or more variables. Sometimes a formula must be rewritten which involves solving for the needed variable. Solving for the needed variable is much like solving general linear equations. The only difference is we will have several variables in the problem and we will need to solve for one specific variable.

Problem 1 :	*Worked Example*

The distance travelled, d, at rate r, for a certain amount of time, t is given by the formula $d = rt$.

a) Solve the formula for t.

Solution.

$$d = rt \quad \text{Isolate } t$$

$$\frac{d}{r} = \frac{rt}{r} \quad \text{Divide both sides of equation by } r$$

$$\frac{d}{r} = t$$

$$t = \frac{d}{r} \quad \text{Our Solution}$$

b) Bert wants to see a concert 60 miles from his home. He expects to travel at a constant speed of 40 mph (miles per hour). How long will Bert be travelling?

Solution.

Since we already solved for t above, we can use that formula to find how long Bert will be travelling by substituing $d = 60$ miles and $r = 40$ mph.

$$t = \frac{d}{r}$$

$$= \frac{60 \, \text{miles}}{40 \, \text{mph}}$$

$$= 1.5 \, \text{hours}$$

It will take Bert 1.5 hours to get to the concert from his home if he travels at a constant speed of 40 mph.

The formula $m + n = p$ is solved for p. If we want to instead solve for n, we need to isolate the n on one side all by itself, with all the other terms on the other side of the equation.

Problem 2 : | *Worked Example*

Solve $m + n = p$ for n.

    Solution.

$$m + n = p \quad \text{Isolate } n$$
$$\underline{-m \quad -m} \quad \text{Subtract } m \text{ from both sides of equation}$$
$$n = p - m \quad \text{Our Solution}$$

As p and m are not like terms, they cannot be combined. For this reason we leave the expression as $p - m$.

It is important to note that we have completed the problem when the variable we are solving for is isolated or alone on one side of the equation and it does not appear anywhere on the other side of the equation.

Problem 3 : | *Media/Class Example*

Solve the following formula for the indicated variable.

a) $A = lw$ for w b) $a - b = c$ for a

Problem 4 : | *You Try*

Solve the following formula for the indicated variable.

a) $x + y + z = 180$ for z b) $C = 2\pi r$ for r

Problem 5 :	*Worked Example*

The perimeter, P, of a rectangle is $P = 2w + 2l$ where w is the width and l is the length of the rectangle.

a) Solve the formula for w.

Solution.

$$P = 2w + 2l \quad \text{Isolate } 2w$$

$$\begin{array}{l} P = 2w + 2l \\ \underline{-2l \quad\quad -2l} \quad \text{Subtract } 2l \text{ from both sides of equation} \\ P - 2l = 2w \end{array}$$

$$\frac{P - 2l}{2} = \frac{2w}{2} \quad \text{Divide both sides by 2}$$

$$\frac{P - 2l}{2} = w$$

$$w = \frac{P - 2l}{2} \quad \text{Our Solution}$$

Note. The solution can be rewritten by dividing **each** term in the numerator by the denominator as follows:

$$w = \frac{P}{2} - \frac{2l}{2}$$

$$= \frac{P}{2} - l$$

It is incorrect to simplify only one term in the numerator (*i.e.* divide $2l$ by 2 but not P to get $w = P - l$).

b) Suppose a rectangle is 8 inches long and has perimeter 40 inches. Find the width of the rectangle.

Solution.

Since we already solved for w above, we can use the formula to find the rectangle's width by substituting $P = 40$ inches and $l = 8$ inches.

$$w = \frac{P - 2l}{2}$$

$$= \frac{(40) - 2(8)}{2}$$

$$= \frac{40 - 16}{2}$$

$$= \frac{24}{2}$$

$$= 12$$

So the rectangle is 12 inches wide.

Problem 6 : | *Media/Class Example*

Given the formula $6x - 3y = 24$.

a) Solve the formula for y.   b) Suppose $x = \frac{2}{3}$, find the value of y.

Problem 7 : | *You Try*

Given the formula $3x + 2y = 12$.

a) Solve the formula for x.   b) Suppose $y = -\frac{3}{2}$, find the value of x.

Problem 8 : | *Worked Example*

Solve $5(x - y) = b$ for x.

Solution.

Method #1: Keep the parenthesis as a group

$$5(x - y) = b \quad \text{The variable } x \text{ is inside the parenthesis,}$$
$$\text{keep } (x - y) \text{ as a group}$$

$$\frac{5(x - y)}{5} = \frac{b}{5} \quad \text{Divide each side by 5}$$

$$x - y = \frac{b}{5}$$

$$x - y = \frac{b}{5} \quad \text{Add } y \text{ to each side of equation}$$
$$\underline{+y \quad +y}$$
$$x = \frac{b}{5} + y \quad \text{Our Solution}$$

Medthod #2: Use Distributive Property

$$5(x - y) = b \quad \text{The variable } x \text{ is inside parenthesis}$$
$$\text{distribute to clear the parenthesis}$$

$$5x - 5y = b$$
$$\underline{+5y \quad +5y} \quad \text{Add } 5y \text{ to each side}$$
$$5x = b + 5y$$

$$\frac{5x}{5} = \frac{b + 5y}{5} \quad \text{Divide each side by 5}$$

$$x = \frac{b + 5y}{5} \quad \text{Our Solution}$$

Note. Be very careful as we isolate x that we do not cancel the 5 on numerator and denominator of the fraction. This is not allowed if there is any addition or subtraction in the fraction.

Note. The answers look different from each other but they are, in fact, the same. The solution $x = \dfrac{b + 5y}{5}$ can be rewritten by dividing **each** term in the numerator by the denominator, as follows: $x = \dfrac{b + 5y}{5} = \dfrac{b}{5} + \dfrac{5y}{5} = \dfrac{b}{5} + y$, giving us the same answer as the first method.

Formulas often have fractions in them and can be solved in much the same way we solve any equations with fractions. First identify the LCD and then multiply each term by the LCD. After we simplify, there will be no more fractions in the problem. We can then solve like any general equation.

**Problem 9 : | *Worked Example*

Solve $\dfrac{a}{3}+\dfrac{b}{4}=c$ for a.

   Solution.

We will use the method where we clear the fraction.

$$\frac{a}{3}+\frac{b}{4}=c \quad \text{Find LCD}=12$$

$$(12)\frac{a}{3}+\frac{b}{4}(12)=c(12) \quad \text{Clear fractions by multiplying each term by 12}$$

$$4a+3b=12c$$

$$\begin{array}{l}4a+3b=12c \quad \text{Subtract } 3b \text{ from both sides}\\ \quad\;\; -3b\;\;-3b\\ \hline 4a=12c-3b\end{array}$$

$$\frac{4a}{4}=\frac{12c-3b}{4} \quad \text{Divide both sides by 4: the coefficient of } a$$

$$a=\frac{12c-3b}{4} \quad \text{Our Solution}$$

Note. The solution can be rewritten by dividing **each** term in the numerator by the denominator as follows: $a=\dfrac{12c}{4}-\dfrac{3b}{4}=3c-\dfrac{3b}{4}$. It is incorrect to simplify only one term in the numerator (i.e. divide 12c by 4 but not 3b to get $a=3c-3b$).

Depending on the context of the problem we may find a formula that uses the same letter, one capital and one lowercase. These represent different values and we must be careful not to combine a capital variable with a lower case variable.

Problem 10 : | *Media/Class Example*

Solve the following formula for the indicated variable.

a) $A = \frac{1}{2}bh$ for b

b) $A = \frac{1}{2}h(a+b)$ for b

Problem 11 : | *You Try*

Solve the following formula for the indicated variable.

a) $E = \frac{mv^2}{2}$ for m

b) $A = \frac{a+b+c}{3}$ for c

Practice Problems: *Formulas*

Solve each of the following equations for the indicated variable.

1. $P=a+b+c$ for a

2. $I=prt$ for t

3. $S=L+2B$ for L

4. $E=mc^2$ for m

5. $P=m(n-c)$ for m

6. $V=lwh$ for w

7. $V=\dfrac{\pi Dn}{12}$ for D

8. $x+5y=3$ for x

9. $ax+b=c$ for x

10. $at-bw=c$ for t

11. $V=\dfrac{1}{3}\pi r^2 h$ for h

12. $A=\dfrac{1}{2}h(a+b)$ for a

13. $5a-7b=4$ for a

14. $q=6(L-p)$ for L

15. $3x+2y=7$ for y

16. $C=\dfrac{5}{9}(F-32)$ for F

17. $A=p+prt$ for r

18. $h=vt-16t^2$ for v

19. $S=\pi rh+\pi r^2$ for h

Additional Problems

20. The angles, x,y,z of a triangle add up to $180°$, that is $x+y+z=180°$.

    a) Solve for y.

    b) Suppose angle, $x=34.6°$ and angle, $z=57.2°$. Find the measure of angle y.

21. The circumference, C, of a circle with radius, r, is $C=2\pi r$.

    a) Solve for r.

    b) If a circle has circumference 32 cm. Find its radius. (Use $\pi\approx3.14$)

22. To find the average, A, of three tests, x, y, z, assuming all tests are weighted equally, we use the formula $A=\dfrac{x+y+z}{3}$.

    a) Solve for z.

    b) In order to pass your math class, you have to average 75. Your first two test scores are 78 and 66. What do you need to score on the third test in order to pass the class?

Rescue Roody!

23. Roody is solving for the height, h, of a pyramid given the volume, $V = \frac{1}{3}bh$ where b is the pyramid's base. This is what Roody did but he was told his answer was not simplified. Roody is confused. Help Roody.

$$A = \frac{1}{3}bh$$

$$\frac{A}{\frac{1}{3}b} = \frac{\frac{1}{3}bh}{\frac{1}{3}b}$$

$$\frac{A}{\frac{1}{3}b} = h$$

24. One of Roody's quiz problems is to use the formula, $p = at - b$ to solve for t. Roody is puzzled as to why he received no credit even though he showed his work. This is his work. Help Rudy understand his mistake.

$$p = at - b$$

$$\frac{p}{a} = \frac{at}{a} - b$$

$$\frac{p}{a} = t - b$$

$$\frac{p}{a} + b = t$$

1.8. APPLICATIONS OF LINEAR EQUATIONS & INEQUALITIES

Objective: Solve application problems by creating and solving a linear equation.

Problem Solving Strategies and Tools (PSST)
When first looking at an application problem (or story problem), it is often helpful to read the entire problem and then read it again more slowly to organize your thoughts.

A) Identify the unknown quantity and select a variable to represent it.

B) Write an equation or inequality that models the relationship between the known and unknown quantities.

C) Solve the equation or inequality. Check for reasonableness of solution.

D) Report the solution.

Problem 1 : | Worked Example

A sofa and a love seat together costs \$442.50. The sofa costs double the love seat. How much do they each cost?

Solution.

Using the Problem Solving Strategies and Tools:

1. *Identify unknown quantity and select a variable to represent it.*
   Let x = cost of a love seat
   Let $2x$ = cost of a sofa (since sofa costs double the love seat)

2. *Write an equation that models the relationship between the known and unknown quantities.*

$$\text{cost of love seat} + \text{cost of sofa} = \$442.50$$
$$x + 2x = \$442.50$$

3. *Solve the equation.*

$$x + 2x = 442.50 \quad \text{cost of love seat} + \text{cost of sofa} = \$442.50$$

$$3x = 442.50 \quad \text{Combine like terms}$$

$$\frac{3x}{3} = \frac{442.50}{3} \quad \text{Isolate the variable, } x \text{ by dividing each side by 3}$$

$$x = 147.50 \quad \text{Our solution (Check reasonableness of solution)}$$

4. *Report the solution.*
   Love seat costs \$147.50 and sofa costs 2(\$147.50) = \$295

Problem 2 : | *Worked Example*

Seargent Piper buys a backpack on sale for $54.85. The sale price is 31% off the original price. What is the original price?

Solution.

Using the Problem Solving Strategies and Tools:

1. *Identify unknown quantity and select a variable to represent it.*
   Let b = original price of backpack

2. *Write an equation that models the relationship between the known and unknown quantities.*

$$\text{Original Price} - \text{Discount} = \text{Sale Price}$$
$$b - 0.31b = \$54.85$$

3. *Solve the equation.*

$$b - 0.31b = 54.85 \quad \text{Original Price} - \text{Discount} = \text{Sale Price}$$

$$0.69b = 54.85 \quad \text{Combine like terms}$$

$$\frac{0.69b}{0.69} = \frac{54.85}{0.69} \quad \text{Isolate the variable, } x \text{ by dividing each side by } 0.79$$

$$b = 79.49 \quad \text{Our solution (Check reasonableness of solution)}$$

4. *Report the solution.*
   The backpack's original cost is $79.49.

Problem 3 : | *You Try*

The Huskies have been playing footballl at the Husky Stadium since 1920. Its seating capacity has been increased by 133.61% to the present capacity of 70,083. What was the seating capacity of Husky Stadium in 1920?

Problem 4 : | *Worked Example*

Elmer is a student in Math 089 and needs to spend at least 2 hours each week working on homework or getting help from a tutor at the Math Center. If he spends 35 minutes on Monday, 1 hour and 10 minutes on Thursday, how much more time does Elmer need to spend that week to meet the minimum requirement?

Solution.

1 hour = 60 minutes   Additional information needed

$2 \text{ hours} \cdot \frac{60 \text{ minutes}}{1 \text{ hour}} = 120 \text{ minutes}$ Convert hours to minutes

1 hour 10 minutes = 60 + 10 = 70 mintues

Use Problem Solving Strategies and Tools.

1. *Identify unknown quantity and select a variable to represent it.*
   Let m = number of minutes needed to meet minimum requirement

2. *Write an inequality that models the relationship between the known and unknown quantities.*
   Minutes spent on Monday + Thursday + additional minutes needed ≥ 120

$$35 + 70 + m \geq 120$$

3. *Solve the inequality.*

$35 + 70 + x \geq 120$ Set up the inequality

$105 + x \geq 120$ Combine like terms

$x \geq 15$ Our Solution (Check for reasonableness of the solution)

4. *Report the solution.*

   Elmer needs to spend at least 15 more minutes at the Math Center to meet the minimum requirement.

Problem 5 : | *You Try*

For the 2014 - 2015 season, the Husky Men's Basketball played 14 home games with an average attendance was 6364. With 2 home games remaining, if the Huskies want to exceed last season's average attendance of 6586, what should the average minimum attendance for the last 2 home games be?

In business, to *break even* means that the profit is zero. This will happen when the *cost(s)* incurred to to produce the items equals what the business *takes in*, which is called the *revenue*.

Problem 6 : | *Media/Class Example*

Marylou is selling cupcakes for 50¢ each. The cost to make one cupcake is 22¢. In order to make the cupcakes attractive, Marylou bought 3 dozen special polka dot baking cups that cost her $2.99. How many cupcakes will Marylou need to sell in order to break even?

Problem 7 : | *Media/Class Example*

The perimeter of a rectangle is 44. The length is 5 less than twice the width. Find the dimensions of the rectangle.

Problem 8 : | *You Try*

The angles of a triangle add up to 180°. The second angle is twice the first while the third angle is three times the first. Find the measurement of each angle.

Problem 9 : | *Media/Class Example*

I just realized that my gas gauge says that I have $\frac{1}{10}$ of a tank of gas left in my car. I stop at a gas station but can only fill 4.8 gallons of gas because I did not have enough cash. At this point, I see that my gas gauge points to the halfway mark. How big is my gas tank?

Practice Problems: *Applications of Linear Equations & Inequalities*

Solve the following word problems. Be sure to follow the problem solving strategies and tools.

1. A bicycle and a bicycle helmet cost $731.25. How much did each cost, if the bicycle cost 8 times as much as the helmet?

2. A triangle is isosceles if two sides are equal. Find the measurement of each side of an isosceles triangle if the third side is 12 inches shorter than either of the two equal sides and the perimeter of the triangel is 60 inches.

3. A barbecue grill is discounted 25% from its original price and is on sale for $169.95. Find the original price.

4. An elevator in an old apartment building has a maximum capacity of 550 lbs. A worker, weighing 220 lbs, has to deliver water to several apartments and wants to haul as many water bottles as the elevator can hold. If each water bottle weighs 42 lbs, what is the maximum number of bottles he can bring into the elevator each time, assuming no other person enters the elevator?

5. An eight ft. board is cut into two pieces. One piece is 2 ft longer than the other. How long are the pieces?

6. Every day, Betty spends $3.85 for an espresso drink. She is thinking about making coffee at home and wonders how quickly she will start to save money. After doing some research, she discovers she can buy an espresso machine for $279 and it would cost 75¢ for each cup. How many cups of coffee will it take before Betty starts saving money by making coffee at home?

7. Seattle wastewater rates for a single family residential customer is $12.27 per 100 cubic feet. According to the Seattle Public Utilities, the typical monthly residential bill is $52.76. At this rate, what is the typical amount of wastewater generated by a single family each month? (Source: seattle.gov)

8. In a room containing 45 students, there were twice as many girls as boys. How many of each were there?

9. A warehouse that stores bags of flour is $\frac{3}{4}$ full. A truck just pulled up and loaded 80 bags of flour for delivery. The warehouse is now $\frac{2}{3}$ full. How many bags of flour can the warehouse store at full capacity?

10. The angles of a triangle sum up to 180°. If the second angle of a triangle is 3 times as large as the first angle and the third angle is 30 degrees more than the first angle. Find the measure of the angles.

11. The perimeter of a rectangle is 152 meters. The width is 22 meters less than the length. Find the length and width.

12. Lou is selling lemonade for $1.25 a cup. The cost to make one cup of lemonade is 27¢ and each plastic cup is 16¢. How many cups will he need to sell to make more than $18.75 to watch a movie and buy snacks?

13. Missoula is a big football town with its football team the University of Montana Grizzlies. On one particular game day, 2 hours before game time, the stadium was $\frac{1}{3}$ full. After 30 minutes, an additional 4200 fans filled the stadium. By this time, the stadium was $\frac{1}{2}$ full. What is the stadium's seating capacity?

14. For superbowl 2015, a scalped ticket was being sold for $14,000. This is a 250% mark-up from the original ticket price. What is the original ticket price?

15. May wants to get an average of 95 so she can earn a 4.0 in her math class. Her final grade is the average of 4 tests. Her test scores so far are: 97, 92 and 98. If all the tests are equally weighted, what minimum score does she need in order to earn a 4.0?

16. Jack and Jim began a business with a capital of $7500. If Jack furnished half as much capital as Jim, how much capital did each furnish?

17. Erica is planning a workshop for new business owners. The fee to rent a room is $2075 plus $90 per hour for the catering and support staff. How many hours can she use the room for the workshop and stay within her budget of $2840?

18. A new Honda Civic Hybrid costs $24,500. The car's value depreciates by $2475 every year, on average. How many years can you keep it, if you want ot be able to sell it for $2225?

19. Homeowners in a particular city pay property taxes at the rate of $1.48 for every $1000 of assessed value of their home. A homeowner paid $777 in property taxes this year. How much is the homeowner's property worth?

20. A friend works two jobs. In the first job, he tutors 16 hours per week at the LOFT and earns $10.50 per hour. In his second job, he earns $20.00 per house to mow someone's lawn. He asks you (his math friend) to calculate many lawns he must mow next week to earn at least $353.00 so he can pay his rent?

Puzzle.

21. Here is a common mathematical puzzle that is used to stump people. Think of a number. Double it, add 6, divide it by 2, and then subtract the number you started with. You should get an answer of 3.

    a) Try thinking of different numbers to see if the puzzle really works.

    b) Why does the answer always come out 3? Work it out algebraically.

22. Create your own puzzle and amaze your friends.

CHAPTER 1 ASSESSMENT

Simplify the following expressions.

1. $4-7(3-5)^2$

2. $6(2k+1)-2(k-5)$

Evaluate each expression using the values given.

3. $pq-(q+5)^2$; use $p=2$, $q=-1$

4. $\dfrac{2n}{4m-n}$; use $m=-1$, $n=-4$

Solve the following equations.

5. $6-w=12$

6. $3x+5=7x-5$

7. $3(2y-1)=6y$

8. $\dfrac{3h+2}{2}=\dfrac{5}{8}+\dfrac{3}{4}h$

9. $|6p-3|=27$

10. $|n+7|=-1$

Solve each inequality. Then graph the solution on a number line and write the solution in interval notation.

11. $-10y<5$

12. $8+5c\leq3c-10$

Solve each equation for the indicated variable.

13. $3y-6x=18$ for y

14. $F=\dfrac{9}{5}C+32$ for C

Solve the following word problems.

15. A chair is discounted 20% from its original price and is on sale for \$675. Find the original cost of the chair.

16. The cost to rent a moving van is \$19.95 plus 59¢. How many miles (to the nearest whole mile) can you go if you can afford to pay \$50?

CHAPTER 2
GRAPHING LINEAR EQUATIONS AND INEQUALITIES

2.1. INTRODUCTION TO GRAPHING

Objective: Work with tables, graphs, and linear equations in two variables; interpret a graph.

Consider the following data on median home prices in Seattle.

Year	2007	2009	2011	2013	2015
Price	$451,000	$412,000	$369,000	$403,000	$471,000

Source: www.zillow.com

To get a better sense of how the median home price varies over time, we can build a graph to represent the behavior of the data. In order to do this, we first need to introduce the Rectangular Coordinate System.

Rectangular Coordinate System

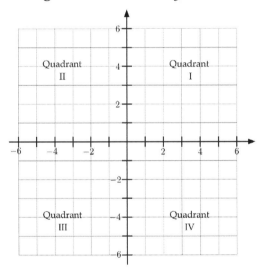

The **Rectangular Coordinate System**, also known as the *xy*-**plane**, is divided into four sections by a horizontal number line (the *x*-**axis**) and a vertical number line (the *y*-**axis**).

These four sections are known as the **quadrants**, which are numbered I, II, III, and IV counterclockwise, as shown on the left. Where the *x*-axis and *y*-axis meet is called the **origin**. This is where $x = 0$ and $y = 0$.

On the *x*-axis, the positive values are located to the right of the origin and negative values are located to the left of the origin. On the *y*-axis, positive values are located above the origin and negative values are located below the origin. We can describe the location of a point in the plane using the horizontal and vertical signed distances the point is from the origin. The first number will be the *x*-**coordinate**, which is the signed distance the point moves left or right from the origin. The second number will be the *y*-**coordinate**, which is the signed distance the point moves up or down from the origin. The location is given as an **ordered pair**, written as (x, y).

World View Note: Locations on the globe are given in a similar manner using latitude and longitude. The origin, located just off the western coast of Africa, is where the prime meridian meets the equator.

81

Problem 1 : | *Worked Example*

Find the ordered pair that describes the location of each point.

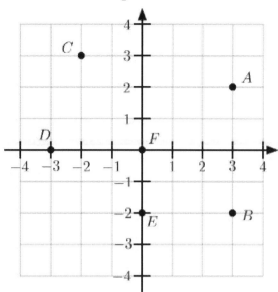

Solution

Starting from the origin, point A is 3 units to the right and 2 units up; we write this as the ordered pair $A(3,2)$.

Likewise, point B is 3 units to the right and 2 units down, so its ordered pair is $B(3,-2)$ and point C is 2 units to the left and 3 units up, so its ordered pair is $C(-2,3)$.

Point D is 3 units directly to the left of the origin; we don't move up or down. Thus, we write $D(-3,0)$ for its ordered pair. Point E is 2 units down from the origin; we don't move left or right so $E(0,-2)$ is its ordered pair. And finally, point F is located at the origin, so $F(0,0)$ is its ordered pair

Note. Order matters with ordered pairs! In the previous example, the points $B(3,-2)$ and $C(-2,3)$ have reversed coordinates and so describe different locations. It's always important to remember that the *x*-coordinate is listed **first** and the *y*-coordinate is listed **second**.

Note. The axes may not use the same scale! As an example, consider the graph below.

Problem 2 : | *Worked Example*

Find the ordered pair that describes the location of each point.

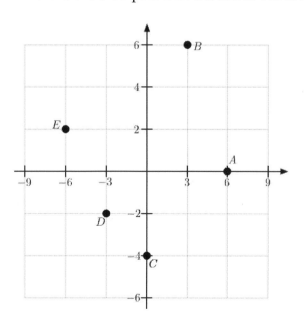

Solution.

Here, each tick mark on the *x*-axis is a multiple of 3 whereas each tick mark on the *x*-axis is a multiple of 2. Because of this, we must take care in labeling points accordingly.

The ordered pairs for the points shown above are

$A(6,0), B(3,6), C(0,-4)$

$D(-3,-2), \text{and } E(-6,2)$.

Problem 3 : | *Class/Media Example*

Find the ordered pair that describes the location of each point, being sure to pay attention to the scale on each axis. Also, state which quadrant each point lies in or whether the point lies on the *x*-axis or *y*-axis.

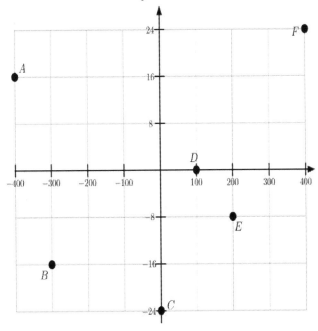

Problem 4 : | *You Try*

Plot the ordered pairs $A(-4, -2), B(3, 8), C(1, -6)$, and $D(0, 6)$. Be sure to pay attention to the scale on each axis.

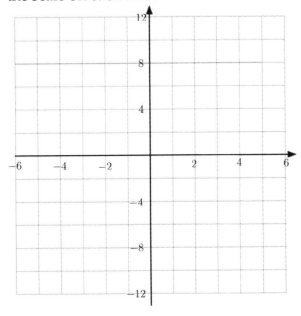

Problem 5 : | *You Try*

Suppose the point (a, b) is located in Quadrant I.

    a) Which Quadrant is the point $(-a, -b)$ located in?

    b) Which Quadrant is the point $(-b, a)$ located in?

Note. Ordered pairs may not always have integer coordinates, such as $(-1.2, 7.4)$ or $\left(\frac{3}{4}, -8\right)$. Because of this, and because of issues with scaling, we may sometimes have to estimate the location when graphing a point.

Problem 6 : | *You Try*

Find the ordered pair that describes the location of each point, being sure to pay attention to the scale on each axis. You may need to estimate the location of some points.

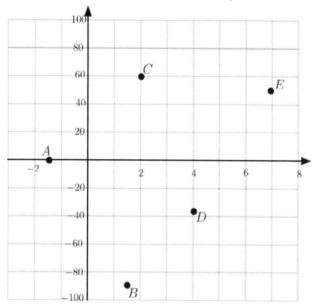

Interpreting Graphs

At the beginning of this section, we looked at the median home price in Seattle by year and mentioned that a graph of the data can help us visualize the behavior of the prices over time. Now that we have established the rectangular coordinate system, we can build such a graph.

Problem 7 : | *Worked Example*

Since we are analyzing how home prices change *over time*, it will be most natural to use the horizontal axis (now labeled the ***t*-axis**, '*t*' for time) for the year and the vertical axis (labeled the ***p*-axis**, '*p*' for price) for the price.

Now, since we have home prices starting in the year 2007, we will make 2007 correspond to $t=0$. Thus, the year 2009 will correspond to $t=2$ since this is 2 years after 2007. The year 2011 will correspond to $t=4$, etc.

Also, since the prices are in the hundreds of thousands, we'll make $100,000 correspond to 100. This then makes the home price $451,000 correspond to 451, the home price $412,000 correspond to 412, etc.

By making these choices, we can consider the entries in the data table as ordered pairs.

Data Table:

Year	Price
2007	$451,000
2009	$412,000
2011	$369,000
2013	$403,000
2015	$471,000

Ordered Pairs:

t	p	Ordered Pair
0	451	(0, 451)
2	412	(2, 412)
4	369	(4, 369)
6	403	(6, 403)
8	471	(8, 471)

We now create a graph of median home prices in Seattle over time by plotting the ordered pairs, being sure to label our axes for clarity. Note that we need to estimate the location of points because of our scale.

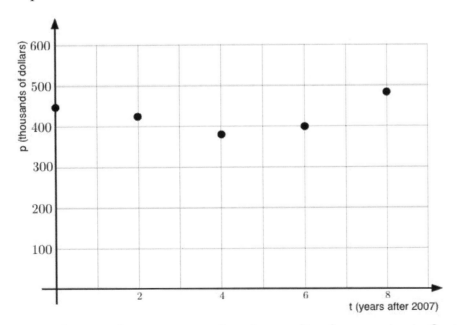

From this graph, we can see that the median home price in Seattle **appears** to decrease from 2007 to 2011 and then begins to increase from 2011 to 2015.

Warning. You must always remember that a data set does not always tell the whole picture of how two quantities are related. This can be due to the fact that our data is not complete (e.g. we do not have data for the years 2008, 2010, etc.). Also, we do not have data for what happens *within* each year, we only have data for the start of each year. For example, our graph above suggests that home prices decrease from 2007 to 2009. However, the price actually *increased* from 2007 to 2008 and then began to decrease (due to the market crash) from 2008 to 2011. A more detailed graph using *semi-annual* prices is shown below. Source: www.zillow.com

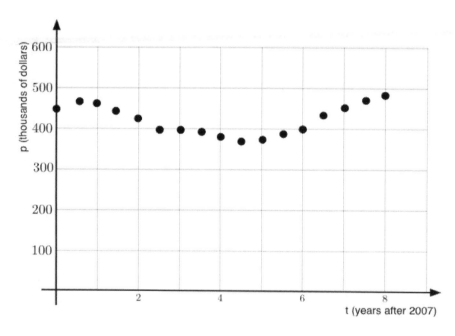

Problem 8 : | *Class/Media Example*

The average price of gasoline in Seattle for selected dates is listed below. (Source: www.seattlegasprices.com)

Date	1/1/2015	3/1/2015	5/1/2015	7/1/2015
Price ($/gallon)	$2.49	$2.85	$3.01	$3.24

a) Interpret the data set as a collection of ordered pairs as we did in the previous example. You will need to choose variables for each axis and state what $t = 0$ corresponds to.

b) Plot the ordered pairs on the axes given. You need to clearly label each axis, indicating appropriate units.

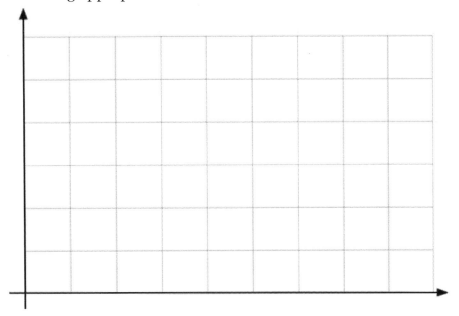

c) What do you observe about the average gas price in Seattle based on the data provided?

Problem 9 : | *Worked Example*

Suppose you are driving your car along a straight road and that your distance from home is shown by the following graph.

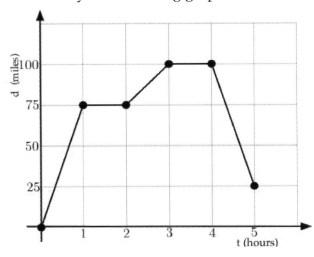

a) Where do you start?

b) When is your car at rest?

c) Find all times when you are driving away from home. Find all times when you are driving towards home.

d) How far from home are you 2.5 hours into your trip?

e) Find all times when you are 100 miles from home.

Solution.

a) We start when $t=0, d=0$. So, you start from home.

b) Your car is at rest when your distance from home is not changing. This happens when $1<t<2$ and when $3<t<4$.

c) You are driving away from home when your distance is increasing. This happens when $0<t<1$ and when $2<t<3$. You are driving towards your home when your distance is decreasing. This happens when $4<t<5$.

d) When $t=2.5, d=87.5$ (midway between 75 and 100). So you are 87.5 miles from home 2.5 hours after your trip began.

e) We are 100 miles from home when $3<t<4$.

Problem 10 : | *Class/Media Example*

Suppose you start hiking along a straight-line path, starting at the trailhead, and your distance from a nearby lake is shown in the graph below.

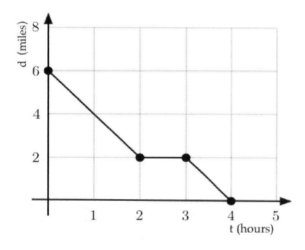

a) How far is the lake from the trailhead?

b) At what time are you 4 miles from the lake?

c) At what time are you 1 mile from the trailhead?

d) You decide to take a break along the way to eat some lunch. When do you do this?

e) After how many hours do you reach the lake?

Graph of an Equation in Two Variables

In Chapter 1, we dealt with equations in one variable, such as $2x-3=5$. The solution to such an equation is a real number ($x=4$ for this equation), and we can plot this on the number line. In this Chapter, we introduce **equations in two variables**, such as $y=4x+7, d=30t$, and $a-b=5$. To specify a solution to this type of equation, we now need to give two real numbers that make the equation true after substitution. So, for example, if we substitute $x=1$ and $y=11$ into the equation $y=4x+7$, we get a true statement since $11=4(1)+7$. Below is a table containing a few other solution pairs to this equation.

Equation: $y=4x+7$

x	y
0	7
-2	-1
3	19
$\dfrac{1}{2}$	9

We can think of each solution pair to this equation as an ordered pair. For the above example, we would then say that the ordered pairs $(1,11), (0,7), (-2,-1), (3,19),$ and $\left(\dfrac{1}{2},9\right)$ are all solutions to the equation.

> A **solution** to an equation in two variables is an ordered pair such that the coordinates of the ordered pair make the equation true after substitution.

Problem 11 : | *Worked Example*

Are the following ordered pairs solutions to the equation $3y-2x=23$?

a) $(-4,5)$

b) $\left(4,\dfrac{5}{3}\right)$

Solution.

a) Substitute $x=-4$ and $y=5$ into the equation to see if the equality holds true.

$$3(5)-2(-4) \stackrel{?}{=} 23$$
$$15+8 = 23\checkmark$$

The ordered pair $(-4,5)$ is a solution to the equation $3y-2x=23$.

b) Substitute $x=4$ and $y=\dfrac{5}{3}$ into the equation to see if the equality holds true.

$$3\left(\dfrac{5}{3}\right)-2(4) \stackrel{?}{=} 23$$
$$5-8 \neq 23$$

The ordered pair $\left(4,\dfrac{5}{3}\right)$ is not a solution to the equation $3y-2x=23$.

Problem 12 : | *Media/Class Example*

Are the following ordered pairs solutions to the equation $y = x^2$?

a) $(2, 4)$ b) $(-2, -4)$

Problem 13 : | *You Try*

Are the following ordered pairs solutions to the equation $y = -\dfrac{2}{3}x + 5$?

a) $(-3, -7)$ b) $\left(-\dfrac{9}{2}, \dfrac{7}{2}\right)$

Practice Problems: *Introduction to Graphing*

Find the ordered pair that describes the location of each point.

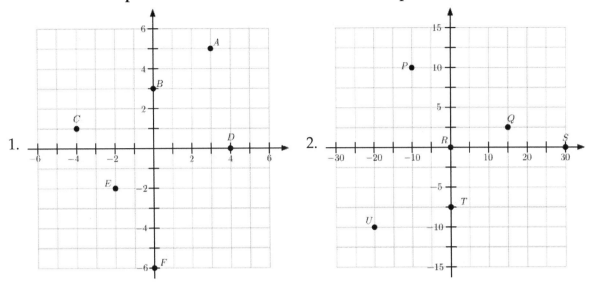

1.

2.

Plot the following ordered pairs and tell what quadrant or axis the point is located.

3. $A(5,3)$

4. $B(-1,3)$

5. $C(4,-2)$

6. $D(0,-7)$

7. $E(-2,-8)$

8. $F(0,3)$

9. $G(-6,0)$

10. $H(1,4)$

11. $I(-5,-2)$

12. $I(9,0)$

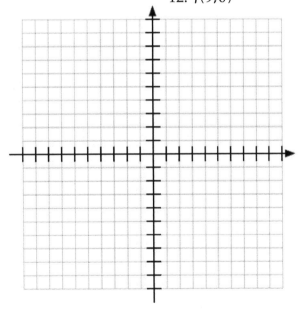

2.2. GRAPHING LINEAR EQUATIONS IN TWO VARIABLES

Objective: To be able to produce a table of solutions to a linear equation in two variables, graph those points and produce a line graph

A **linear equation in 2 variables** can be written as $Ax + By = C$, where A, B, and C are any real numbers. As you learned in Section 2.1, the solutions to this type of equations are ordered pairs, (x, y) whose values make the equation true after substitution.

We can visualize the solutions to an equation in two variables by plotting the ordered pairs. For example, consider the equation $y = 2x$ and the given table of solutions. Plotting the ordered pair solutions and connecting the points with a line gives us the **graph** of the equation.

Equation   *Table of Solutions*   *Graph*

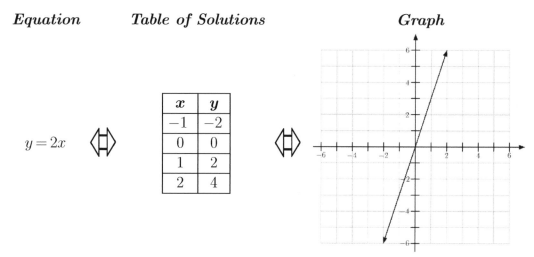

$y = 2x$

x	y
-1	-2
0	0
1	2
2	4

How many solutions are there to a linear equation in 2 variables? An infinite number. We can list some of them in a table; however, it also helps to be able to see them on a graph.

In fact the relationship between two variables can be expressed in any of these three ways: equation, table of solutions or graph. Ultimately, you will be able to start with any one of these and produce either of the other two.

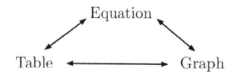

Problem 1 : | *Media/Class Example*

Given the graph of the equation $y = -x + 4$, produce a table of 3 solutions by reading the points off the graph.

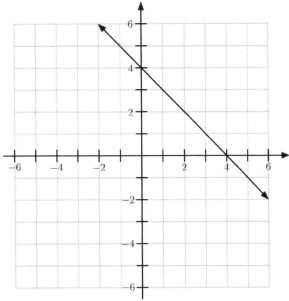

Then, confirm that each ordered pair solves the equation.

Confirm solution 1 ordered pair (__ , __):

Confirm solution 2 ordered pair (__ , __):

Confirm solution 3 ordered pair (__ , __):

The graph of all the solutions of a linear equation in two variables is a straight line. Our goal in this section is to start with an equation, produce a table of solutions, and then graph the line.

Problem 2 : | *Worked Example*

Graph the equation $2x + y = 1$.

> **Solution.**

We start by producing a table of at least three solutions. Two points define a line but the third point ensures that your line is correct. We substitute a value for either the x- or y-coordinate and then solve for the other coordinate. It does not matter what value we choose to start with, though some choices end up being easier to graph.

Let us begin with $x = 1$. What is the y-coordinate of the solution? Substitute 1 for x and solve for y:

$$2(1) + y = 1$$
$$2 + y = 1 \quad \text{Subtract 2 from each side}$$
$$y = -1$$

Our solution to this equation is $(1, -1)$.

Next, let's substitute $y = -2$. What is the x-coordinate of the solution?

$$2x + (-2) = 1 \quad \text{Add 2 to each side}$$
$$2x = 3 \quad \text{Divide each side by 2}$$
$$x = \frac{3}{2}$$

Another solution to our equation is $\left(\frac{3}{2}, -2\right)$. Fractions are not always easy to graph, so we will look for another point.

Let $y = 5$, solve for x:

$$2x + 5 = 1 \quad \text{Subtract 5 from each side}$$
$$2x = -4 \quad \text{Divide each side by 2}$$
$$x = -2$$

A third solution is $(-2, 5)$.

Finally, let $x = -1$ and solve for y:

$$2(-1) + y = 1$$
$$-2 + y = 1 \quad \text{Add 2 to each side}$$
$$y = 3$$

A fourth solution is $(-1, 3)$.

Let's put these solutions into a table.

Equation

$$2x + y = 1$$

Solutions

x	y
-2	5
-1	3
1	-1
$\dfrac{3}{2}$	-2

Now that we have at least 3 points, we can graph the line that represents the equation.

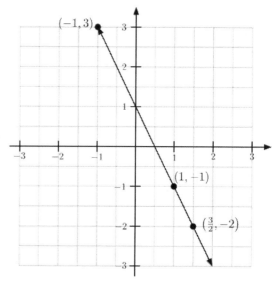

Problem 3 : *Media/Class Example*

Graph the equation $x - 2y = 4$ by first producing a table of at least 3 solutions.

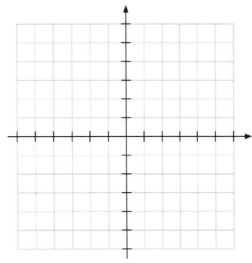

x	y

Note. You can also manipulate the linear equation and solve for y.

Problem 4 : *Media/Class Example*

Graph the equation $y = 3x$ by first producing a table of at least 3 solutions.

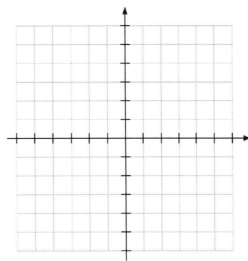

x	y

Problem 5 : | *You Try*

For each equation, produce a table of at least 3 solutions and then graph the line.
a) $y = 2x - 3$

x	y

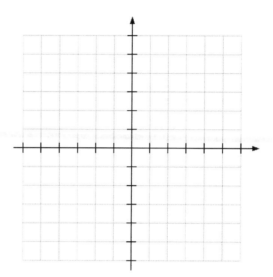

b) $x - y = 0$

x	y

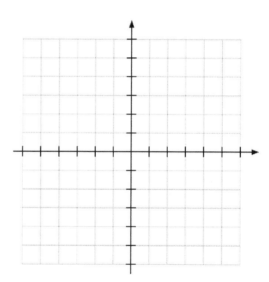

World View Note: When French mathematicians Rene Descartes and Pierre de Fermat first developed the coordinate plane and the idea of graphing lines (and other functions) the y-axis was not a vertical line!

Intercepts

The **y-intercept** of a line is the point where the line crosses the y-axis. What is the x-coordinate of any y-intercept?

The **x-intercept** of a line is the point where the line crosses the x-axis. What is the y-coordinate of any x-intercept?

Problem 6 : | *Worked Example*

Here is the graph for the equation $y = x + 4$. Notice the labeled intercepts.
y-intercept: $(0, 4)$ and x-intercept: $(-4, 0)$

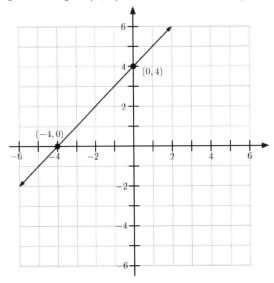

Problem 7 : | *Media/Class Example*

From the graph, write the x-intercept and y-intercept as ordered pairs.

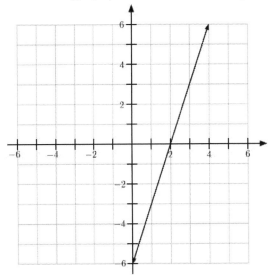

Problem 8 : | *You Try*

For each graph, write the x-intercept and y-intercept as ordered pairs.

a)

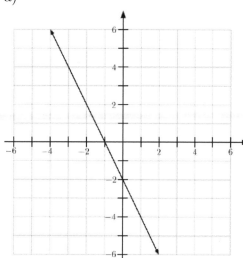

b)

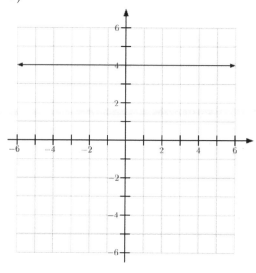

> **Given an equation, how do we find the intercepts?**
> To find the x-intercept, substitute 0 for y and solve for x.
> To find the y-intercept, substitute 0 for x and solve for y.

Problem 9 : | *Worked Example*

Find the intercepts of the equation $x + 2y = -4$.

   Solution.

To find the x-intercpet, we substitute 0 for y and solve for x.

$$\begin{aligned} x + 2y &= -4 \\ x + 2(0) &= -4 \\ x + 0 &= -4 \\ x &= -4 \end{aligned}$$

The x-intercept is the point $(-4, 0)$.
To find the y-intercept, we substitute 0 for x and solve for y.

$$\begin{aligned} x + 2y &= -4 \\ 0 + 2y &= -4 \\ 2y &= -4 \\ y &= -2 \end{aligned}$$

The y-intercept is the point $(0, -2)$.

Problem 10 : | *Media/Class Example*

For each equation, find the x-intercept and y-intercept. Write your answers as ordered pairs.

a) $y = \dfrac{1}{2}x + 3$

b) $y = -3x$

Problem 11 : | *You Try*

For each equation, find the x-intercept and y-intercept. Write your answers as ordered pairs.

a) $y = 2x - 3$

b) $x - y = 0$

You can use the intercepts to graph the equation of a line.

Problem 12 : | *You Try*

For each equation, find the x-intercept and y-intercept. Then graph the line using those points.

a) $y = \dfrac{1}{2}x - 3$

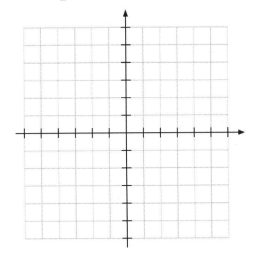

b) $3x + y = 6$

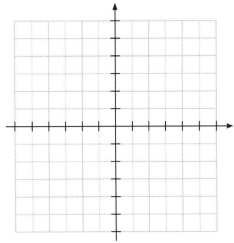

c) $5x - y = 5$

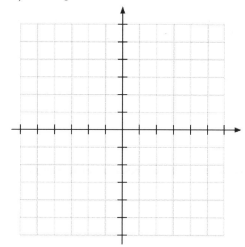

d) $x + y = -3$

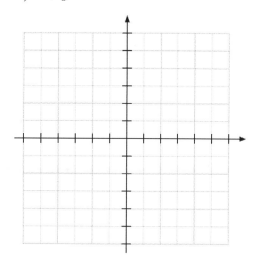

The formula to convert from Celsius to Fahrenheit is $F = \dfrac{9}{5}C + 32$. Find the intercepts and interpret the result.

Vertical and Horizontal Lines

Consider the following linear equation in *two* variables:

$$y = 2.$$

The first question you might ask is "Where is the other variable?" Remember the form of a linear equation in 2 variables, $Ax + By = C$. In this equation, $A = 0$.

Problem 14 :	Worked Example

Graph $y = 2$.

Solution.

Let us start by building a table.

Equation   Solutions

$y = 2$

x	y
-1	?
0	?
1	?
2	?

When $x = -1$, what is y?
Since there is no x term in the equation,
there is no place to substitute. Our answer is $y = 2$.

What about when $x = 0$? Same answer, $y = 2$.

The pattern we are seeing here is that whatever the x-coordinate is, the y-coordinate of the solution must be $y = 2$.

Equation   **Solutions**

$y = 2$

x	y
-1	2
0	2
1	2
2	2

What does the graph of $y = 2$ look like? Let's plot the points and sketch the line.

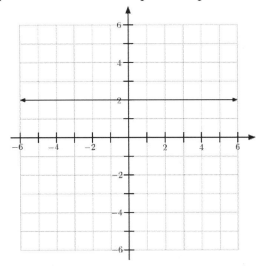

Problem 15 : | *Media/Class Example*

Graph the line $x = -3$.

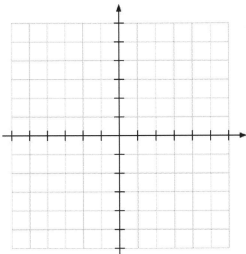

Let us summarize what we have seen:

- A line where all the y-coordinates are equal is a **horizontal line** and it's equation takes the form $y = $ a real number.
- A line where all the x-coordinates are is a **vertical line** and it's equation takes the form $x = $ a real number.

Problem 16 : | *You Try*

Graph the following lines.

a) $y = -1$

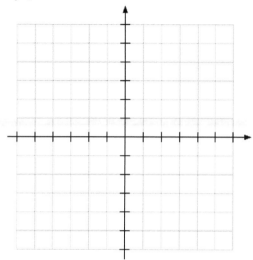

c) $x = 0$

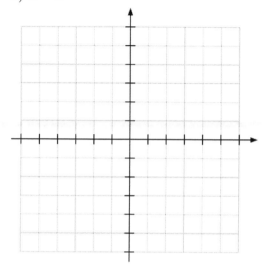

b) $x = 2$

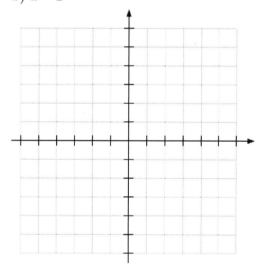

d) $y = 0$

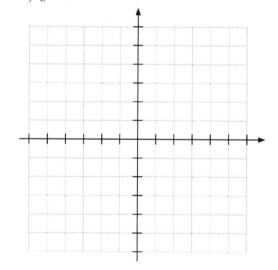

Practice Problems: *Graphing Linear equations in two variables*

For each graph, identify the *x*- and *y*-intercepts, if they exist. Write your answer as ordered pairs.

1.

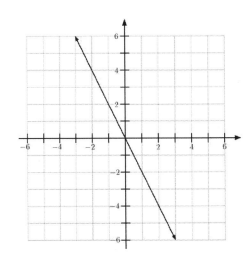

4.

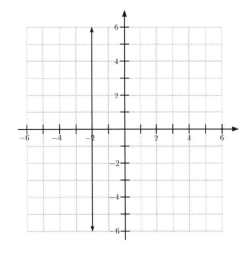

2.

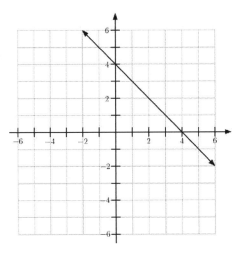

5.

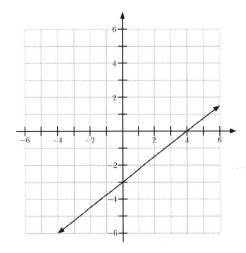

3.

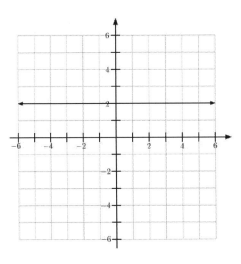

6.

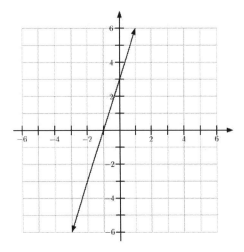

For each equation, find the x- and y-intercepts, if they exist. Then sketch the line, finding more points as necessary.

7. $y = -\dfrac{1}{4}x - 3$

8. $y = x - 1$

9. $y = x$

10. $y + 4x = 2$

11. $y = -1$

12. $x - y = -3$

13. $x = 1$

14. $3x + y = 0$

15. $4x + y = 5$

16. $3x + 5y = 15$

17. $y = \dfrac{1}{2}x$

18. $y = -x$

19. $-3x + y = 6$

20. $x - 2y = 8$

MID-CHAPTER CHECK-UP

1. Is the following ordered pair $\left(-\dfrac{1}{3}, \dfrac{20}{7}\right)$ a solution to the equation $7y - 3x = 21$?

2. For each equation, produce a table of at least 3 solutions. Then using those points, graph the line.

    a) $3x - 2y = 6$

    b) $y = -3$

3. For each equation, find the x-intercept and y-intercept. Write your answer as an ordered pair. Then graph the line using those points.

    a) $y = -x + 5$

    b) $2x + y = -4$

2.3. Slope

Objective: Understand the meaning of the slope of a line, find the slope given two points and from the graph of a line, and use the slope formula to solve rate of change applications.

What do these descriptions have in common?

- pitch or steepness of a roof

- grade of a road on a mountain

- incline of a wheelchair ramp

These all refer to the steepness of the surface. **Slope** is the measure of steepness.
One of the interesting properties of a line is its slope. A line with a large positive slope, like 25, is very steep. A line with a small positive slope, such as $\frac{1}{20}$, is close to flat. We also use slope to describe the direction of a line.

- A line that goes up from left to right has a **positive slope**.

- A line that goes down from left to right has a **negative slope.**

| **Problem 1 :** *Class/Media Example* |

Determine whether the line has a positive or a negative slope.

a)   b)   c)

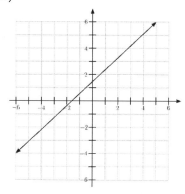   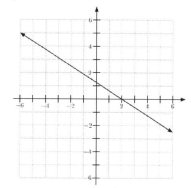   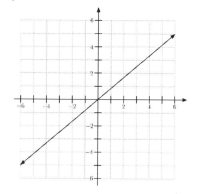

Reading Slope from a Graph
As we measure steepness, we are interested in how fast the line rises compared to how far the line runs.

- **rise** is the vertical change or the change in the y-coordinates

- **run** is the horizontal change or the change in the x-coordinates

We can write this as an equation: $\text{slope} = \dfrac{\text{rise}}{\text{run}} = \dfrac{\text{change in } y}{\text{change in } x}$

To compute the slope of a line

- Pick two points on the line. It is easier to compute slope if the x and y-coordinate of the point are both integers.

- Count the number of units it takes to go from one point to the other point in the direction of the coordinate axes.

  For the rise, the vertical change or change in y is positive when you go up and negative when you go down.

  For the run, the horizontal change or change in x is positive when you go to the right and negative when you go left.

Problem 2 : | *Worked Example*

Find the slope of the given line.   **Solution**

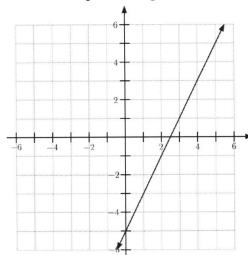
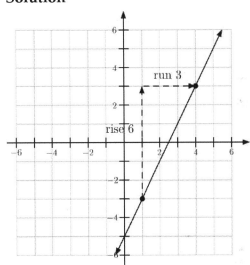

Note that the line goes up from left to right, so we should use a positive slope.

Pick two points on the line. From one point to the next, the graph goes 6 units up and 3 units right. This is a rise of 6 and a run of 3.

So the slope written as a fracction is $\dfrac{\text{rise}}{\text{run}} = \dfrac{6}{3}$, which simplifies to 2.

Our solution: slope is 2

Problem 3 : | *Worked Example*

Find the slope of the given line.   **Solution:**

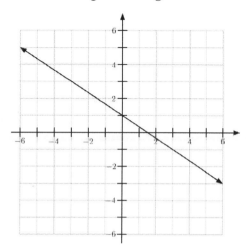   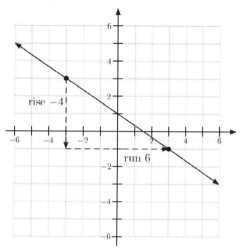

Note that the line goes down from left to right. We should have a negative slope. Pick two points on the line. From one point to the next, the graph goes down 4, and right 6. This is a rise of −4, and a run of 6.

Slope is written as a fraction, $\dfrac{\text{rise}}{\text{run}} = \dfrac{-4}{6}$ or $-\dfrac{2}{3}$.

Our solution: slope is $-\dfrac{2}{3}$.

Problem 4 : *Class/Media Example*

Choose two points on the line and use them to find the slope of each line

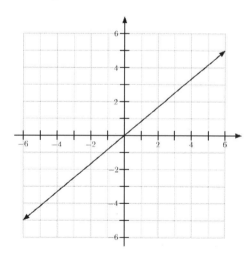   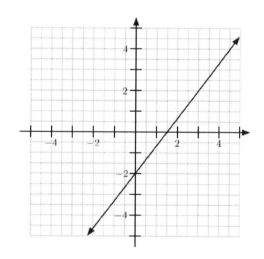

Problem 5 : | *You Try*

Choose two points on the line and use them to find the slope of each line.

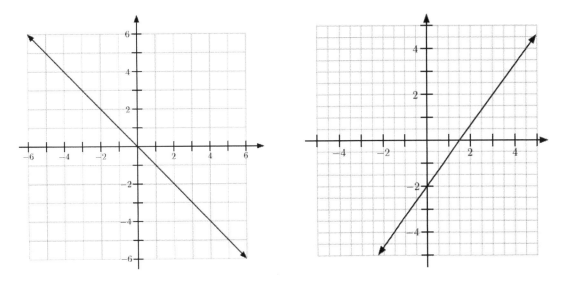

Which two points did you choose? Notice that *any* two points on a line can determine the slope and the slope will be the same between all points on the line.

Horizontal and Vertical Lines

We need to be aware of two special kind of lines that have unique slopes. Recall from Section 2.2 that a horizontal line contains points with the same y-coordinate. What does that mean regarding slope? Given two points on a horizontal line, the rise will always be 0 for any run or slope $= \dfrac{0}{\text{run}}$. Therefore, a horizontal line has slope $= 0$.

Similarly, a vertical line contains points with the same x-coordinate. Given two points on a vertical line, for any rise, the run will always be 0 or slope $= \dfrac{\text{rise}}{0}$. However, dividing by 0 is not possible. Therefore, a vertical line has undefined slope.

Problem 6 : | *Worked Example*

Find the slope of the following lines.

a)

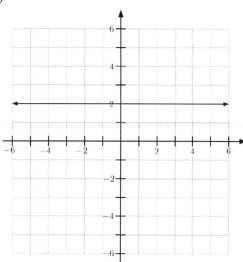

b)

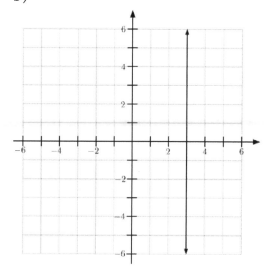

Solution.

a)

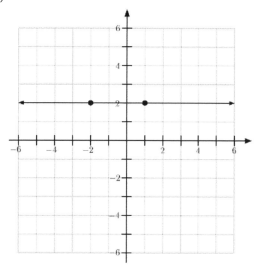

b)

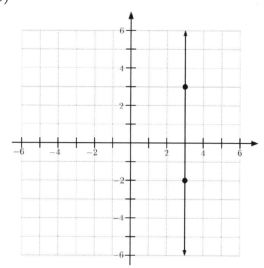

This line has no rise but the run is 3 units. Slope is $\frac{0}{3} = 0$. This line and all horizontal lines have 0 slope.

This line has a rise of 5, but no run. Slope is $\frac{5}{0}$ which is undefined. This line and all vertical lines have undefined slope.

As you can see, there is a big difference between having a zero slope and having an undefined slope. Remember, slope is a measure of steepness. A horizontal line is not steep at all. In fact, it is flat. Therefore, a horizontal line has zero slope.

On the other hand, a vertical line cannot get any steeper. It is so steep that there is no number large enough to express how steep it is. Therefore, a vertical line has an undefined slope.

To summarize:

> - A horizontal line has slope=0.
>
> - A vertical line has undefined slope.

Problem 7 : | *You Try*

Find the slope of the following lines.

a)

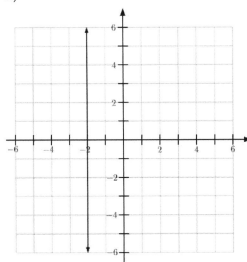

b)

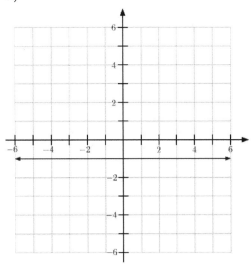

Slope Formula

In addition to reading a graph, we can find the slope of a line through any two points using the **slope formula**. Given any two points (x_1, y_1) and (x_2, y_2), the slope of the line containing those points is

$$\text{Slope}=m=\frac{y_2-y_1}{x_2-x_1}$$

Problem 8 : | *Worked Example*

Find the slope of the line containing the points $(-4, 3)$ and $(2, -9)$.

   Solution.

Let $(x_1, y_1) = (-4, 3)$ and $(x_2, y_2) = (2, -9)$. Then use the slope formula to find the slope of the line containing the points.

$$m = \frac{y_2-y_1}{x_2-x_1}$$

$$= \frac{-9-3}{2-(-4)}$$

$$= \frac{-12}{6}$$

$$= -2$$

What if we let $(x_1, y_1) = (2, -9)$ and $(x_2, y_2) = (-4, 3)$. Let us compute the slope of the line.

$$m = \frac{y_2 - y_1}{x_2 - x_1}$$

$$= \frac{3 - (-9)}{-4 - 2}$$

$$= \frac{12}{-6}$$

$$= -2$$

We can see that it does not matter which point we call (x_1, y_1) and which point we call (x_2, y_2). The slope of the line containing the points will still be the same.

Problem 9 : | *Class/Media Example*

Find the slope of the line containing the following points.

a) $(4, 6)$ and $(2, -1)$ b) $(-4, -1)$ and $(-4, -5)$

Problem 10 : | *You Try*

Find the slope of the line containing the following points.

a) $(3, 1)$ and $(-2, 1)$ b) $(6, -5)$ and $(-2, 7)$

Using the slope formula, we can also find missing coordinates if we know what the slope is.

Problem 11 : | *Worked Example*

A line with slope -3 passes through $(2, y)$ and $(5, -1)$. Find the value of y.

Solution.

$$m = \frac{y_2 - y_1}{x_2 - x_1} \quad \text{Use the slope formula}$$

$$-3 = \frac{-1 - y}{5 - 2} \quad \text{Simplify}$$

$$-3 = \frac{-1 - y}{3} \quad \text{Multiply each side by 3}$$

$$-9 = -1 - y \quad \text{Add 1 to each side}$$

$$-8 = -y \quad \text{Divide each side by} -1$$

$$8 = y \quad \text{Our Solution}$$

Problem 12 : | *You Try*

A line with slope $\frac{2}{5}$ passes through $(-3, 2)$ and $(x, 6)$. Find the value of x.

We can also use a point and a slope value to find another point.

Problem 13 : | *Worked Example*

Given the point $(2, 1)$, find two other points that create a slope of $\frac{4}{3}$.

   Solution.

First, let us rewrite the slope as a fraction, $m = \dfrac{\text{rise}}{\text{run}} = \dfrac{4}{3} = \dfrac{-4}{-3}$. This means that for every rise of 4, we need a corresponding run of 3. It also means that for every rise of -4, we need a corresponding run of -3.

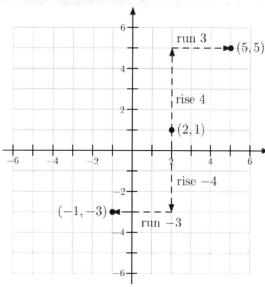

Starting at point $(2, 1)$, if we go up 4 and right 3, we arrive at the point $(4, 5)$.

Starting at point $(2, 1)$, if we go down 3 and left 2, we arrive at the point $(0, -2)$.

Therefore, two other points that create a slope of $\frac{3}{2}$ if we start at the point $(2, 1)$ are $(3, 4)$ and $(0 - 2)$.

Problem 14 : | *Class/Media Example*

Given the point $(1, 0)$, find two other points that make a slope of -3.

Problem 15 : | *You Try*

a) Given the point $(-3, 2)$, find two other points that make a slope of $-\frac{5}{2}$.

b) Given the point $(2, 1)$, find two other points that make a slope of 0.

Rate of Change and Applications

Another way to describe slope is **Rate of Change**. In the context of an application, it is important to include the units when providing an answer.

Problem 16 : | *Worked Example*

On a drive from Seattle, WA to San Francisco, CA, you notice that after 2 hours, you had driven 124 miles. After 5 hours, you had driven 310 miles. What is the rate of change of distance with respect to time?

Solution.

Let us start by writing the given data as 2 points. Since we want rate of change of distance with respect to time, we select time as the first coordinate and miles as the second coordinate. The 2 points are: $(2, 124)$ and $(5, 310)$.

Next, substitute the points into the slope formula.

$$m = \frac{y_2 - y_1}{x_2 - x_1}$$

$$= \frac{310 - 124}{5 - 2}$$

$$= \frac{186}{3} = 62$$

What are the units? The y-coordinates are miles and the x-coordinates are hours. So our final answer is 62 miles/hour.

Problem 17 : | *Class/Media Example*

At the start of 2012, the clarity of Lake Tahoe was measured to a depth of 75.3 feet. In 1999, the clarity of the lake was 66 feet. What is the rate of change of the depth of clarity with respect to time? Be sure to include units in your answer.

(Source: http://news.ucdavis.edu/search/news_detail.lasso?id=11171)

Problem 18 : | *You Try*

According to the Americans with Disabilities Act (ADA), a wheelchair ramp must have a slope no steeper than 1:12 (or $\frac{1}{12}$). If the ramp is required for an 8 inch high curb, how far away from the curb must the ramp end?
(Source: www.access-board.gov/guidelines-and-standards/buildings-and-sites/about-the-ada-standards/ada-standards)

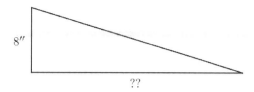

Direct Variation

Consider the drive from Seattle to San Francisco again. Assume we have several data points.

Time (hour)	Distance (mile)
1	62
2	124
3	186
4	248
5	310

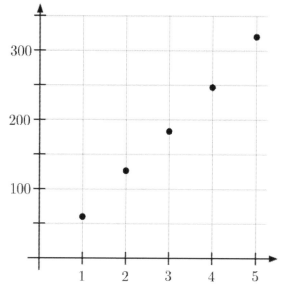

We know that the rate of change or slope between $(2, 124)$ and $(5, 310)$ is 62 mph. Does this rate of change hold for all the other combination of points as well?

Let us try the points $(1, 62)$ and $(2, 124)$ and substitute them in the slope formula.

$$m = \frac{y_2 - y_1}{x_2 - x_1}$$

$$= \frac{124 - 62}{2 - 1}$$

$$= \frac{62}{1} \text{ or } 62 \text{ mph}$$

In fact, we see that for every increase of 1 hour, the distance traveled increased by 62 miles. This tells us that distance varies directly with time. In other words, distance is a constant multiple of time. This is known as **direct variation.** The multiple, which is the slope, is known as the **constant of variation.**

Problem 19 : | You Try

a) Assuming that the distance varies directly with time on our drive from Seattle to San Francisco. How far will we have gone after 8 hours?

b) It is approximately 800 miles from Seattle to San Francisco. How many hours will the complete drive take?

Practice Problems: *Slope*

Find the slope of each line.

1.

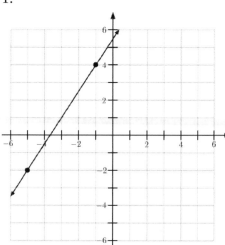

2.

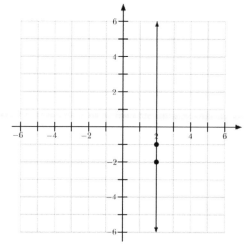

3.

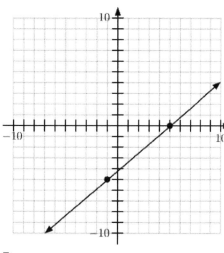

4.

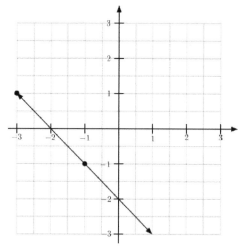

5)

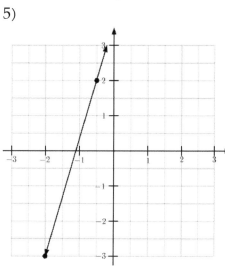

6)

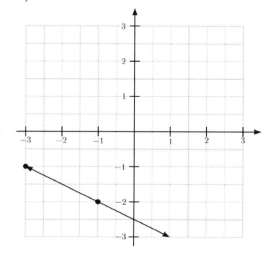

7)

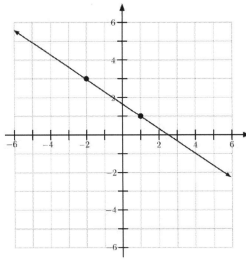

8)

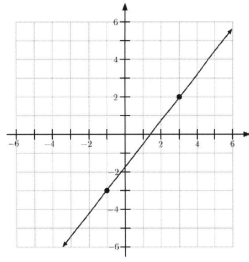

9)

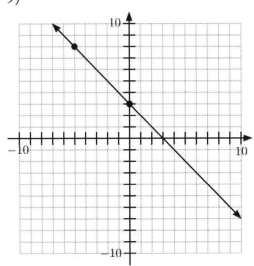

10)

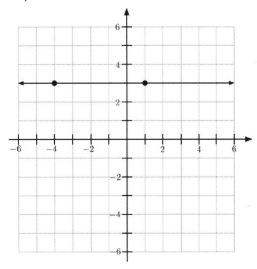

Find the slope of the line through each pair of points.

11. $(-2, 10), (-2, -15)$

12. $(1, 2), (-6, -14)$

13. $(-15, 10), (16, -7)$

14. $(13, -2), (7, 7)$

15. $(10, 18), (-11, -10)$

16. $(-3, 6), (-20, 13)$

17. $(-16, -14), (11, -14)$

18. $(13, 15), (2, 10)$

19. $(-4, 14), (-16, 8)$

20. $(9, -6), (-7, -7)$

Find the value of x or y so that the line through the points has the given slope.

21. $(-8, y)$ and $(-1, 1)$; slope: $\dfrac{6}{7}$

22. $(8, y)$ and $(-2, 4)$; slope: $-\dfrac{1}{5}$

23. $(-2, y)$ and $(2, 4)$; slope: $\dfrac{1}{4}$

24. $(2, -5)$ and $(3, y)$; slope: 6

Using the given point, find another point that makes the provided slope.

25. $(6, 14)$; $m = \dfrac{1}{2}$

26. $(5, -2)$; undefined slope

27. $(-3, 8)$; $m = -4$

28. $(0, 2)$; $m = 0$

29. $(-4, 7)$; $m = 5$

30. $(1, 0)$; $m = -\dfrac{3}{2}$

Applications

31. The recommended dosage of a certain medication for a 50 lb. dog is 25 mL. For a 16 lb. dog, the dosage recommendation is 8 mL.

   a) What is the rate of change of the dosage with respect to weight?

   b) What would be the recommended dosage for a 35 lb. dog?

32. In January 2012, the average 2 bedroom home in Seattle was worth \$308,000. In January 2015, that value was \$410,000. What is the rate of change of home prices with respect to time? (Source: www.zillow.com/home-values/)

33. In 2004, Washington State recycled or diverted 6,223,974 tons of waste. In 2013, Washington State recycled or diverted 7,961,040 tons. What is the rate of change of recyled or diverted waste with respect to time?   (Source: www.ecy.wa.gov/beyondwaste/bwprog_swGenRec.html)

34   **Rescue Roody!**

Roody needs to find the slope between these points: $(3, 4)$ and $(1, -2)$. He cannot seem to find the correct slope. Help him determine what he did wrong.

$$m = \frac{y_2 - y_1}{x_2 - x_1}$$

$$= \frac{-2 - 4}{3 - 1}$$

$$= \frac{-6}{2} \quad \text{or} \quad -3$$

Challenge
For each table, determine if the data represents a direct linear variation or not.

a)

x	y
2	3
5	8
11	18
14	23

b)

x	y
1	1
2	4
3	9
4	16

2.4. SLOPE-INTERCEPT FORM

Objective: Understand the slope-intercept form and use it to graph a line.

If we know the slope of a line, m, and we know a point on that line, (x_1, y_1), then we can use the slope formula to determine all other points that belong to the line: a point (x, y) will belong to this line if it makes a slope of m with the point (x_1, y_1). In this section we will work with the case where the point is the y-intercept, so it is of the form $(x_1, 0)$.

Example 1

A line has slope $m = \dfrac{2}{3}$ and it's y-intercept is at $(0, 1)$. Find the equation of the line.

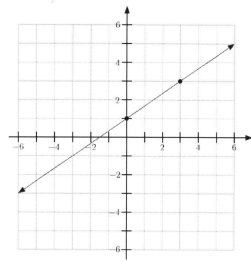

A point (x, y) will be on the line if it makes a slope of $m = \dfrac{2}{3}$ with the point $(0, 1)$.

Using the slope formula and simplifying we obtain

$$\frac{y-1}{x-0} = \frac{2}{3} \qquad \text{slope formula}$$

$$\frac{y-1}{x} = \frac{2}{3} \qquad \text{simplify}$$

$$\frac{y-1}{x}(x) = \frac{2}{3}(x) \qquad \text{multiply by } x \text{ each side}$$

$$y - 1 = \frac{2}{3}x$$

We have found our equation. If we solve for y we obtain

$$y = \frac{2}{3}x + 1$$

and from here we can notice that

- the coefficient of x is the slope of the line.
- the y-intercept can be easily read: using $x = 0$ we obtain:

$$y = \frac{2}{3}(0) + 1$$

$$y = 0 + 1$$

$$y = 1$$

Summarizing we can see that for the equation $y = \frac{2}{3}x + 1$:

- the slope is $m = \frac{2}{3}$

- the y-intercept is $(0, 1)$

This will happen every time we have an equation of a line in this form.

The equation $y = mx + b$

- has slope m

- has y-intercept at the point $(0, b)$.

This is called the **slope-intercept** form of the equation of a line.

Problem 1 : | *Worked example*

Identify the slope and the y-intercept of the line $y = 2x + 3$.

    Solution.

This equation is in slope-intercept form, where $m = 2$ and $b = 3$. The slope of the line is $m = 2$ and the y-intercept is the point $(0, 3)$.

Problem 2 : | *Class/Media Example*

Identify the slope and the y-intercept of the line $y = \frac{2}{3}x - 2$.

Problem 3 : | *You Try*

Identify the slope and the y-intercept of the line $y = 4x + 13$.

Problem 4 : | *Worked Example*

Identify the slope and the y-intercept of the line $4x - 3y = 9$.

   Solution.

This equation is not in slope-intercept form. We must solve for y first:

$$4x - 3y = 9$$
$$\underline{-4x \qquad = -4x} \qquad \text{subtract } 4x \text{ from each side}$$
$$-3y = -4x + 9$$

$$\frac{-3y}{-3} = \frac{-4x + 9}{-3} \qquad \text{divide each side by (-3)}$$

$$y = \frac{-4x}{-3} + \frac{9}{-3} \qquad \text{simplify}$$

$$y = \frac{4x}{3} - 3 \qquad \text{our equation in slope-intercept form}$$

Now that we have the equation in slope intercept form we can see that the slope is $m = \frac{4}{3}$ and the y-intercept is $(0, -3)$.

Problem 5 : | *Class/Media Example*

Identify the slope and the y-intercept of the line $3x + 6y = 9$.

Problem 6 : | *You Try*

Identify the slope and the y-intercept of the line $4x + 2y = -7$.

Problem 7 : | *Worked example*

Identify the slope and the y-intercept of the line $3x - 2y = 10$. Use this information to graph the line.

Solution.

This line is not written in slope-intercept form. We must solve for y first:

$$3x - 2y = 10$$
$$\underline{-3x = -3x} \qquad \text{subtract } 3x \text{ from each side}$$
$$-2y = -3x + 10$$

$$\frac{-2y}{-2} = \frac{-3x + 10}{-2} \qquad \text{divide each side by } (-2)$$

$$y = \frac{-3x}{-2} + \frac{10}{-2} \qquad \text{simplify}$$

$$y = \frac{3x}{2} - 5$$

So the slope of the line is $m = \dfrac{3}{2}$ and the y-intercept is $(0, -5)$. We can now graph this line. The steps are shown below.

-Locate the y-intercept: $(0, -5)$

-Starting at $(0, 5)$ we move 2 units in the x-direction and 3 units in the y-direction.

The point $(2, -2)$ is on the line.

We connect these two points and continue the line in each direction.

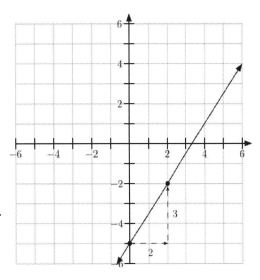

Problem 8 : | *Class/Media example*

Identify the slope and the y-intercept of the equation $2x - 3y = 6$. Use the information to sketch the graph of the line.

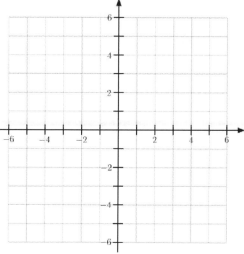

Problem 9 : | *You Try*

Identify the slope and the y-intercept of the linear equation $4x - 3y = 9$. Use the information to sketch the graph of the line.

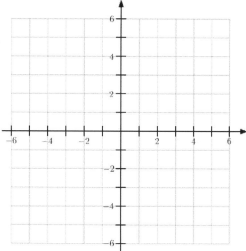

When graphing lines with this method we have to be careful with the sign of the slope. Starting from our y-intercept, to obtain a second point on the line we have two cases:

- If the slope is positive, say $m = \dfrac{c}{d}$, we can move in the positive x-direction and positive y-direction from our y-intercept. But we can also move in the negative direction for both, since $\dfrac{-c}{-d} = \dfrac{c}{d}$.

- If the slope is negative, say $m = -\dfrac{c}{d}$, we can move in the positive x-direction and **negative** y-direction from our y-intercept, since $-\dfrac{c}{d} = \dfrac{-c}{d}$. But we can also move in the negative x-direction and positive y-direction, since since $-\dfrac{c}{d} = \dfrac{c}{-d}$.

Problem 10 : | Worked Example

Identify the slope and the y-intercept of the linear equation $y = -\dfrac{4}{3}x + 3$. Use the information to sketch the graph of the line.

Solution.

This equation is given in slope-intercept form. The slope is $m = \dfrac{-4}{3}$ and the y-intercept is $(0, 3)$.

-Locate the y-intercept: $(0, 3)$

-Starting at $(0, 3)$ we move 3 units in the x-direction and 4 units in the **negative** y-direction.

The point $(3, -1)$ is on the line.

We connect these two points and continue the line in each direction.

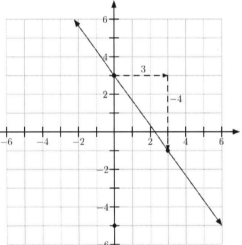

Problem 11 : | *Class/media example*

Identify the slope and the *y*-intercept of the linear equation $2x + 3y = 12$. Use the information to sketch the graph of the line.

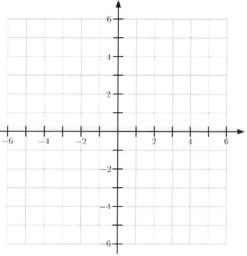

Problem 12 : | *You try*

Identify the slope and the *y*-intercept of the linear equation $5x + 3y = 15$. Use the information to sketch the graph of the line.

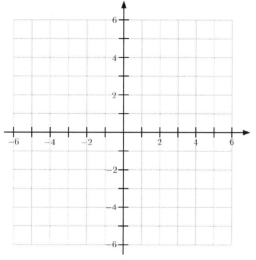

Problem 13 : | *Worked example*

Find the equation of the line given below. Write your answer in slope-intercept form.

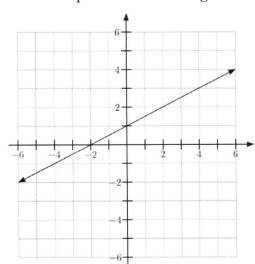

We can read from the graph that the y-intercept is at $(0,1)$ and that a second point on the graph is $(4,3)$.

We can find the slope of the line by using these two points,

$$m = \frac{3-1}{4-0} = \frac{2}{4} = \frac{1}{2}$$

So the equation of the line is

$$y = \frac{1}{2}x + 1$$

Problem 14 : | *Class/Media example*

Find the equation of the line given below. Write your answer in slope-intercept form.

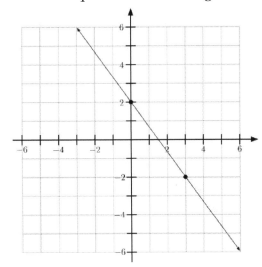

Problem 15 : | *You-try*

Find the equation of the line given below. Write your answer in slope-intercept form.

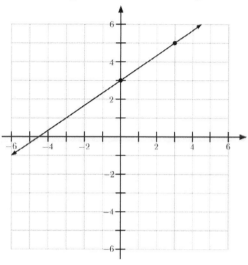

Problem 16 : | *Worked example*

Find the equation of the line given below. Write your answer in slope-intercept form.

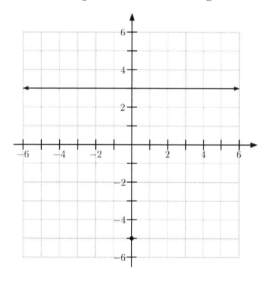

We can read from the graph that the y-intercept is at $(0,3)$ and that a second pointon the graph is $(4,3)$.

The slope of this line is $m=0$.

So the equation of the line is

$$y=0x+3$$

or simply

$$y=3$$

Problem 17 : | *Worked example*

Find the equation of the line given below. Write your answer in slope-intercept form, if possible.

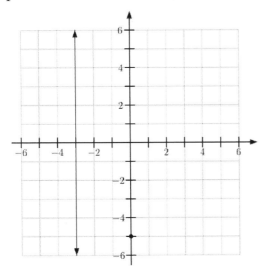

If we want to calculate the slope of this line we can choose the points $(-3,1)$ and $(-3,4)$ and obtain

$$m = \frac{4-1}{-3-(-3)} = \frac{3}{0}$$

which is undefined.

We can notice however that any point on this line will have -3 as the x-value, that is, any point of the form $(-3, y)$ will belong to this line.

The equation of this line is

$$x = -3$$

Problem 18 : | *Worked example*

You-Drive, a moving truck rental company charges a \$20.00 initial fee plus \$0.90 per mile driven. Write an equation that relates the number of miles driven with the total cost of renting a truck.

Solution Let's start by making a table of values where x is the number of miles driven and y is the total cost of renting a truck with this company.

x	y
0	20.00
1	20.90
2	21.80
3	22.70
4	23.60
5	24.50

These points are on a line where the slope is $m = 0.90$ and the y-intercept is $(0, 20)$, so the equation is $y = 0.90x + 20$.

Problem 19 : | *Class/Media Example*

A bowling alley charges $3 for shoe rental, and $25 per hour. Tom and 3 of his friends are planning to go bowling this weekend. Write an equation that relates the number of hours they play with the amount each one of them will pay.

Problem 20 : | *You Try*

A College student-club wants to sell t-shirts to generate funds for their expenses. A print-shop charges $30 for the design, plus $7.50 to print each t-shirt. Write an equation that relates the number of t-shirts printed with the total cost of printing them.

Problem 21 : | *You Try*

In Seattle a taxi charges a $3.40 drop charge (applied as soon as you enter the taxi) plus a distance charge of $3.40 per mile. Write an equation that relates the amount of miles of a taxi ride witht the total amount paid.

Practice Problems: *Slope-Intercept Form*

Write the slope-intercept form of the equation of each line given the slope and the *y*-intercept.

1) Slope = 2, *y*-intercept = (0,5)

2) Slope = −6, *y*-intercept = (0,4)

3) Slope = 1, *y*-intercept = (0,−4)

4) Slope = −1, *y*-intercept = (0,−2)

5) Slope = $-\dfrac{3}{4}$, *y*-intercept = (0,−1)

6) Slope = $-\dfrac{1}{4}$, *y*-intercept = (0,3)

7) Slope = $\dfrac{1}{3}$, *y*-intercept = (0,1)

8) Slope = $\dfrac{2}{5}$, *y*-intercept =(0,5)

Write the slope-intercept form of the equation of each line.

9)

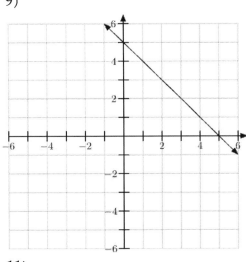

10)

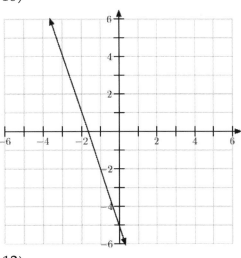

11)

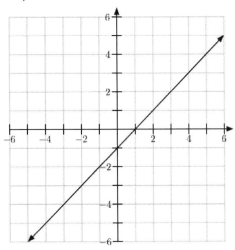

12)

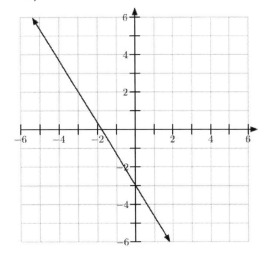

13)

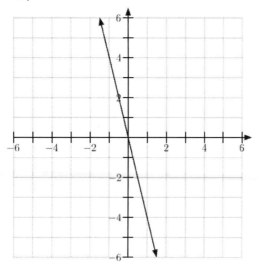

14)

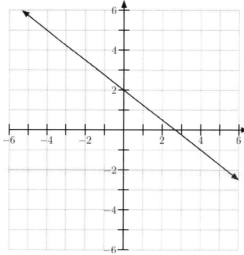

15) $x + 10y = -37$

16) $x - 10y = 3$

17) $2x + y = -1$

18) $6x - 11y = -70$

19) $7x - 3y = 24$

20) $4x + 7y = 28$

21) $x = -8$

22) $x - 7y = -42$

23) $y - 4 = -(x + 5)$

24) $y - 5 = \dfrac{5}{2}(x - 2)$

25) $y - 4 = 4(x - 1)$

26) $y - 3 = -\dfrac{2}{3}(x + 3)$

27) $y + 5 = -4(x - 2)$

28) $0 = x - 4$

29) $y + 1 = -\dfrac{1}{2}(x - 4)$

30) $y + 2 = \dfrac{6}{5}(x + 5)$

Sketch the graph of each line.

31) $y = \dfrac{1}{3}x + 4$

32) $y = -\dfrac{x}{5} - 4$

33) $y = \dfrac{6}{5}x - 5$

34) $y = -\dfrac{3}{2}x - 1$

35) $2y = 3x$

36) $x + 5y + 20 = 0$

37) $x - y + 3 = 0$

38) $4x + 5 = 5y$

39) $y + 4 - 3x = 0$

40) $2y - 8 = 6x$

41) $-3y = -5x + 9$

41) $x + y = 0$

2.5. POINT-SLOPE FORM

Objective: Use point-slope form to write the equation of a line.

If we know the slope of the line, and we know the y-intercept, we can easily write the equation of the line. However, there are many occasions when we do not know the y-interccept, but we know a different point on the line. If that is the case, we have another useful formula for finding the equation of the line.

If we let the slope of an equation be m, a specific point (given) on the line be (x_1, y_1) and any another point on the line be (x, y); we can use the slope formula to obtain:

$$\frac{y - y_1}{x - x_1} = m \quad \text{Slope formula; multiply each side by } (x - x_1)$$

$$y - y_1 = m(x - x_1) \quad \text{Point slope form}$$

$$\boxed{\text{Point-Slope Form: } y - y_1 = m(x - x_1)}$$

Problem 1 : | Worked Example

Write the equation of the line through the point $(3, -4)$ with a slope of $\frac{3}{5}$.

Solution.

$$y - y_1 = m(x - x_1) \quad \text{Insert values into point slope form}$$

$$y - (-4) = \frac{3}{5}(x - 3) \quad \text{Simplify signs}$$

$$y + 4 = \frac{3}{5}(x - 3) \quad \text{Our Solution}$$

We often prefer final answers written in slope-intercept form. If the directions ask for the answer in slope-intercept form, we will distribute the slope, then solve for y.

Problem 2 : | Worked Example

Write the slope-intercept form of the equation of the line that goes through the point $(-6, 2)$ and has a slope of $-\frac{2}{3}$.

Solution.

$$y - y_1 = m(x - x_1) \quad \text{Insert values into point slope formula}$$

$$y - 2 = -\frac{2}{3}(x - (-6)) \quad \text{Simplify signs}$$

$$y - 2 = -\frac{2}{3}(x + 6) \quad \text{Distribute slope}$$

$$y - 2 = -\frac{2}{3}x - 4 \quad \text{Solve for } y \text{ by adding 2 to each side of equation}$$

$$y = -\frac{2}{3}x - 2 \quad \text{Our Solution}$$

Note. Be very careful with signs when using the point-slope formula.

Problem 3 :	*Media/Class Example*

Write the slope-intercept form of the equation of the line through the given point and given slope.

    a) Through the point $(4, -3)$ with slope -2

    b) Through the point $(-1, 2)$ with slope $-\dfrac{3}{2}$

    c) Through the point $(5, 8)$ with slope 0

Problem 4 :	*You Try*

Write the slope-intercept form of the equation of the line through the given point and given slope.

    a) Through $(-1, 1)$ with slope 4

b) Through $(-1,-4)$ with slope $-\dfrac{2}{3}$

c) Through $(6,5)$ with undefined slope

If we are given two points but not the slope, we can also find the equation of the line. However, we will first need to find the slope.

Problem 5 : | *Worked Example*

Find the equation of the line through the points $(-2,5)$ and $(4,-3)$.

    Solution.

$$m=\frac{y_2-y_1}{x_2-x_1} \quad \text{First we must find the slope}$$

$$m=\frac{-3-5}{4-(-2)}=\frac{-8}{6}=-\frac{4}{3} \quad \text{Insert the values in the slope formula and evaluate}$$

$$y-y_1=m(x-x_1) \quad \text{With the slope and one point, use the point slope form}$$

$$y-5=-\frac{4}{3}(x-(-2)) \quad \text{Simplify}$$

$$y-5=-\frac{4}{3}(x+2) \quad \text{Our Solution}$$

Problem 6 : | *Worked Example*

Find the equation of the line through the points $(-3, 4)$ and $(-1, -2)$ in slope-intercept form.

Solution.

$$m = \frac{y_2 - y_1}{x_2 - x_1} \quad \text{First we must find the slope}$$

$$m = \frac{-2-4}{-1-(-3)} = \frac{-6}{2} = -3 \quad \text{Insert the values in slope formula}$$

$$y - y_1 = m(x - x_1) \quad \text{With slope and either point, use the point slope formula}$$

$$y - 4 = -3(x - (-3)) \quad \text{Simplify}$$

$$y - 4 = -3(x + 3) \quad \text{Distribute}$$

$$y - 4 = -3x - 9 \quad \text{Solve for } y \text{ by adding 4 to each side}$$

$$y = -3x - 5 \quad \text{Our Solution}$$

Problem 7 : | *Worked Example*

Find the equation of the line through the points $(6, -2)$ and $(-4, 1)$ in slope-intercept form.

Solution.

$$m = \frac{y_2 - y_1}{x_2 - x_1} \quad \text{First we must find the slope}$$

$$m = \frac{1-(-2)}{-4-6} = \frac{3}{-10} = -\frac{3}{10} \quad \text{Insert values into slope formula}$$

$$y - y_1 = m(x - x_1) \quad \text{Use slope and either point, use point slope formula}$$

$$y - (-2) = -\frac{3}{10}(x - 6) \quad \text{Simplify signs}$$

$$y + 2 = -\frac{3}{10}(x - 6) \quad \text{Distribute slope}$$

$$y + 2 = -\frac{3}{10}x + \frac{9}{5} \quad \text{Solve for } y \text{ by subtracting 2 from each side}$$

$$y = -\frac{3}{10}x - \frac{1}{5} \quad \text{Our Solution}$$

Problem 8 : | *Media/Class Example*

Write the slope-intercept form of the equation of the line through the given points.

a) $(1,3)$ and $(-3,3)$ b) $(-1,-4)$ and $(5,0)$

Problem 9 : | *You Try*

Write the slope-intercept form of the equation of the line through the given points.

a) $(0,2)$ and $(2,4)$ b) $(-5,-1)$ and $(5,-2)$

Problem 10 : | *Media/Class Example*

Wrtie the slope-intercept form of the equation of the given line.

a)   b)

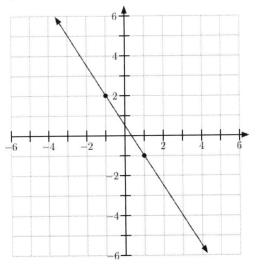   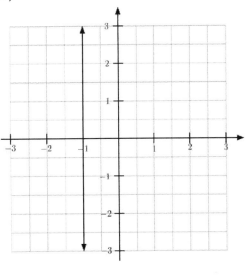

Problem 11 : | You Try

Wrtie the slope-intercept form of the equation of the given line.

a)

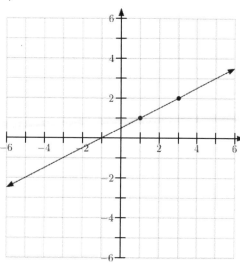

b)

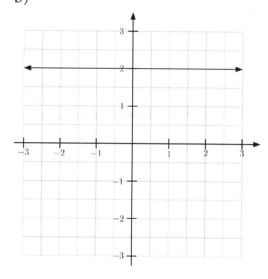

Problem 12 : | Media/Class Example

The World Health Organization (WHO) has been monitoring the percent of population without improved drinking water source. Five years after they began their study, 20% of the world population were without improved drinking water source. Ten years after they began their study, 17% of the world population were without improved drinking water source. Let y be the percent of world population without improved drinking water source x years after WHO began their study. (Source: WHO/UNICEF Joint Monitoring Programme for Water Supply and Sanitation).

a) Find two ordered pairs that represent the given information.

b) Identify the slope and interpret the slope in the context of the situation.

c) Write a linear equation, in slope-intercept form, that represents this situation.

d) What is the y-intercept? Interpret the meaning of the y-intercept in the context of the situation.

e) If the trend continues, what percent of the population would you expect to be without improved drinking water source 12 years after WHO began their study?

f) Approximately how many years after the study began did 14% of the world population not have improved drinking water source?

Problem 13 : | *You Try*

Tuition for the academic year 2015-2016 for Washington State residents taking classes at North Seattle College increases at the same rate for the first 10 credits. If you take 3 credits, tuition is $308.25. If you take 5 credits, tuition is $513.75. Let y represent tuition for x credits taken.

a) Write the ordered pairs that represent the above information.

b) Identify the slope.

c) Interpret the slope in the context of the situation.

d) Write a linear equation, in slope-intercept form, that represents the relationship between the number of credits and tuition.

e) Suppose you want to take 10 credits of classes, how much will your tuition be?

Problem 14 : | *You Try*

Lee is a heavy smoker, smoking 2 packs a day (20 cigarettes in a pack) but he is determined to quit smoking. He is in a new program that will help him quit. The goal is to cut back 3 cigarettes each week. Let x be the number of weeks Lee has been on the program and y be the number of cigarettes smoked per week.

a) Write a linear equation, in slope-intercept form, that represents the situation above.

b) Identify the y-intercept.

c) Interpret the meaning of the y-intercept in the context of the situation.

d) If Lee follows the program, how many weeks will it take before he quits smoking entirely.

Practice Problems: *Point Slope Form*

Write the point-slope form of the equation of the line through the given point with the given slope.

1. through $(-1,-5)$, slope $=9$

2. through $(2,2)$, slope $=\frac{1}{2}$

3. through $(0,-7)$, slope $=-\frac{1}{4}$

4. through $(-4,1)$, slope $=0$

5. through $(2,3)$, slope is undefined

6. through $(-1,0)$, slope $=-\frac{5}{4}$

7. through: $(-4,3)$ and $(-3,1)$

8. through: $(1,6)$ and $(-3,6)$

9. through: $(-1,-4)$ and $(-5,0)$

10. through: $(-8,1)$ and $(-8,4)$

11. through: $(-4,-2)$ and $(0,4)$

12. through: $(3,5)$ and $(-5,3)$

Write the slope-intercept form of the equation of the line through the given point with the given slope.

13. through: $(2,-2)$, slope $=1$

14. through: $(-1,-7)$, slope $=2$

15. through: $(3,4)$, undefined slope

16. through: $(4,0)$, slope $=-\frac{3}{2}$

17. through: $\left(-\frac{1}{2},-\frac{3}{4}\right)$, slope $=0$

18. through: $(-2,-2)$, slope $=-\frac{2}{3}$

19. through: $(0,2)$ and $(5,-3)$

20. through: $(0,1)$ and $(-3,0)$

21. through: $(4,1)$ and $(1,4)$

22. through: $\left(\frac{2}{7},-1\right)$ and $\left(\frac{2}{7},-2\right)$

23. through: $(-5,1)$ and $(-1,-2)$

24. through: $(1,-1)$ and $(-5,-4)$

Write the slope-intercept form of the equation of the line shown.

25.

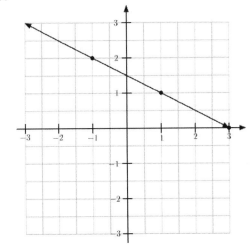

26.   27.

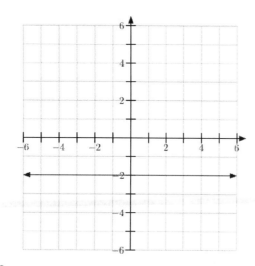
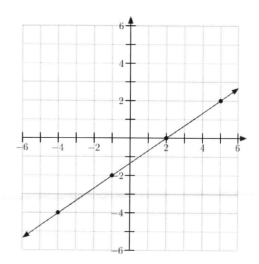

28.

Applications.

29. Elmer, who weighs 250 lbs., joined a diet program that promises to help him lose 2 pounds each week or his money back.

    a) Write a linear equation, in slope-intercept form, that represents the situation above. Let y represent Elmer's weight after being on the diet program for x weeks.

    b) After 4.5 weeks, how much does Elmer weigh if he follows the diet program?

    c) If Elmer sticks with the program, how many weeks will it take before he gets to his target weight of 175 lbs?

30. A yogurt company recently released a new yogurt blend. The company expects to make $68,000 in profit at the end of the first year. At the end of the 4th year, the company expects profit to be $272,000. Let y represent profit at the end of x years.

    a) Write the ordered pairs that represent the above information.

    b) Identify the slope and interpret the slope in the context of the situation.

    c) Write a linear equation, in slope-intercept form, that represents the relationship between year and profit.

    d) Use the equation to find the company's profit at the end of the 6th year.

31. For the academic year 2015-2016, tuition for International students taking classes towards a BAS degree at North Seattle College increases at the same rate for the first 10 credits. If you take 2 credits, tuition is $1,197.68. If you take 7 credits, tuition is $4,191.88. Let y represent tuition for x credits taken.

    a) Write the ordered pairs that represent the above information.

    b) Identify the slope and interpret the slope in the context of the situation.

    c) Write a linear equation, in slope-intercept form, that represents the relationship between the number of credits and tuition.

    d) Suppose an international student wants to take 10 credits of classes, how much will the tuition be?

32. Cars usually depreciate linearly. You bought a Kia Soul. A year later, the value of your Kia is \$17,530. Three years later, the value of your Kia is \$11,849. Let y be the value of your Kia x years after you purchased your car.

    a) Find two ordered pairs that represent the given information.

    b) Identify the slope.

    c) Interpret the slope in the context of the situation.

    d) Write a linear equation, in slope-intercept form, that represents the relationship between the number of years your own the Kia and its value.

    e) What is the y-intercept?

    f) Interpret the meaning of the y-intercept in the context of the situation.

2.6. PARALLEL AND PERPENDICULAR LINES

Objective: Identify the equation of a line given a parallel or perpendicular line.

Problem 1 : | *You Try*

Each graph shows a pair of parallel lines.

    a) Find the slope of each line.

    b) What did you discover?

i)   ii)

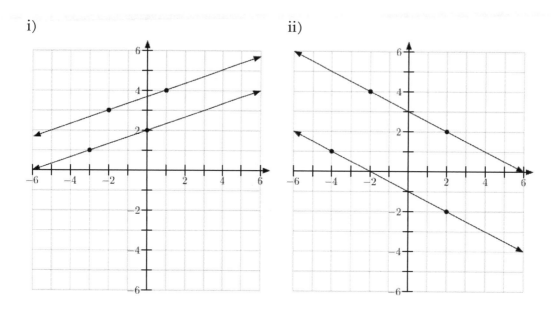

Problem 2 : | *You Try*

Each graph shows a pair of perpendicular lines.

  a) Find the slope of each line.

  b) What did you discover?

i)

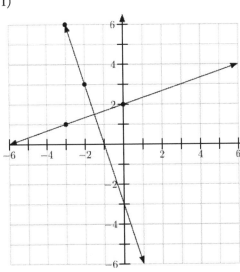

ii)

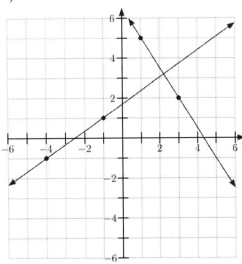

As the above graphs illustrate, parallel lines have the same slope and perpendicular line have slopes that are negative reciprocals of one another.

Problem 3 : | *Worked Example*

Find the slope of a line parallel to $5y - 2x = 7$.

  Solution.

$$5y - 2x = 7 \quad \text{To find the slope, we will put}$$
$$\text{the equation in slope-intercept form}$$

$$5y - 2x = 7$$
$$\underline{+2x + 2x} \quad \text{Add } 2x \text{ to each side}$$
$$5y = 2x + 7$$

$$\frac{5y}{5} = \frac{2x}{5} + \frac{7}{5} \quad \text{Divide each term by 5}$$

$$y = \frac{2}{5}x + \frac{7}{5} \quad \text{Slope Intercept Form}$$

$$m = \frac{2}{5} \quad \text{Slope of the given line}$$

A parallel line has the same slope as the given line

$$m = \frac{2}{5} \quad \text{Slope of parallel line}$$

World View Note: Greek Mathematician Euclid lived around 300 BC and published a book titled, *The Elements*. In it is the famous parallel postulate which mathematicians have tried for years to drop from the list of postulates. The attempts have failed, yet all the work done has developed new types of geometries!

| **Problem 4 :** | *Worked Example* |

Find the slope of a line perpendicular to $3x - 4y = 2$.

Solution.

$$3x - 4y = 2 \quad \text{To find the slope we will re-write the equation in slope-intercept form}$$

$$\begin{aligned} 3x - 4y &= 2 \\ \underline{-3x \quad\quad -3x} & \quad \text{Subtract 3x from each side} \\ -4y &= -3x + 2 \quad \text{Put the } x \text{ term first after the equal sign} \end{aligned}$$

$$\frac{-4y}{-4} = \frac{-3x}{-4} + \frac{2}{-4} \quad \text{Divide each term by } -4$$

$$y = \frac{3}{4}x - \frac{1}{2} \quad \text{Slope intercept form}$$

$$m = \frac{3}{4} \quad \text{Slope of the given line}$$

The slope of a perpendicular line is the negative reciprocal of the slope of the given line

$$m = -\frac{4}{3} \quad \text{Slope of the perpendicular line}$$

Problem 5 : | *Media/Class Example*

a) Find the slope of a line parallel to $6x - 5y = 20$.

b) Find the slope of a line perpendicular to $8x + 3y = -9$.

Problem 6 : | *You Try*

a) Find the slope of a line parallel to $3x + 4y = -8$.

b) Find the slope of a line perpendicular to $y - 3y = 3$.

Problem 7 : | *You Try*

Determine if the following pairs of lines are parallel, perpendicular, or neither.

a)   b)

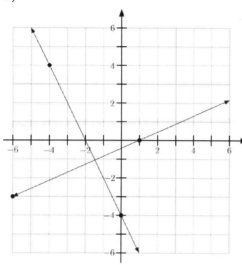   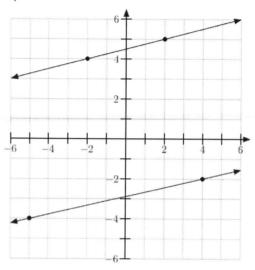

When we have a slope and one point on the second line, it is possible to find the equation of the second line.

Problem 8 : | *Worked Example*

Find the equation of a line through $(4, -5)$ and parallel to $2x - 3y = 6$. Write the line is slope-intercept form.

   Solution.

$$2x - 3y = 6 \quad \text{Write the equation in slope intercept form}$$

$$\begin{array}{r} 2x - 3y = 6 \\ \underline{-2x \quad\quad -2x} \\ -3y = -2x + 6 \end{array} \quad \text{Subtract 2x from each side}$$

$$\frac{-3y}{-3} = \frac{-2x}{-3} + \frac{6}{-3} \quad \text{Divide each term by -3}$$

$$y = \frac{2}{3}x - 2 \quad \text{Slope Intercept Form}$$

$$m = \frac{2}{3} \quad \text{Slope of given line}$$

$$m = \frac{2}{3} \quad \text{Slope of parallel line}$$

$$y - y_1 = m(x - x_1) \quad \text{Substitute } m = \frac{2}{3} \text{ and the given point } (4, -5)$$

$$y - (-5) = \frac{2}{3}(x - 4) \quad \text{Simplify signs}$$

$$y + 5 = \frac{2}{3}(x - 4)$$

$$y + 5 = \frac{2}{3}x - \frac{8}{3} \quad \text{Distribute}$$

$$y = \frac{2}{3}x - \frac{8}{3} - \frac{15}{3} \quad \text{subtract } 5, \left(\frac{15}{3}\right) \text{ from each side}$$

$$y = \frac{2}{3}x - \frac{23}{3} \quad \text{Our solution}$$

Problem 9 : | *Worked Example*

Find the equation, in slope-intercept form, of the line through $(6, -9)$ perpendicular to $y = -\frac{3}{5}x + 4$.

Solution.

$$y = -\frac{3}{5}x + 4 \quad \text{Identify the slope, coefficient of } x$$

$$m = -\frac{3}{5} \quad \text{Slope of given line}$$

$$m = \frac{5}{3} \quad \text{Slope of perpendicular line}$$

$$y - y_1 = m(x - x_1) \quad \text{Substitute } m = \frac{5}{3} \text{ and the given point } (6, -9)$$

$$y - (-9) = \frac{5}{3}(x - 6) \quad \text{Simplify signs}$$

$$y + 9 = \frac{5}{3}(x - 6) \quad \text{Distribute the slope}$$

$$y + 9 = \frac{5}{3}x - 10 \quad \text{Solve for } y \text{ by subtracting 9 from each side}$$

$$y = \frac{5}{3}x - 19 \quad \text{Our Solution}$$

Special Case

A horizontal line is perpendicular to a vertical line. Since a vertical line has undefined slope, we do not look for a negative reciprocal slope. Sketching both lines is often helpful.

| Problem 10 : | *Worked Example* |

Find the equation of the line through $(3, 4)$ perpendicular to $x = -2$.

Solution.

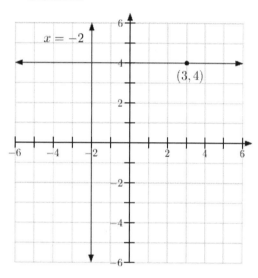

The line $x = -2$ is a vertical line.

The line we are looking for must be a horizontal line.

Since it goes through the point $(3, 4)$, it is the line $y = 4$.

| Problem 11 : | *Media/Class Example* |

Find the equation of the line.

a) through the point $(5, -2)$ and parallel to $y = \frac{3}{5}x$

b) through the point $(5, -1)$ and perpendicular to $5x - 2y = -10$

c) through the point $(2, 3)$ and parallel to $y = 4$

Problem 12 : | You Try

Find the equation of each line.

a) through the point $(-4, 1)$ and parallel to $4x + 3y = -9$

b) through the point $(5, 2)$ and perpendicular to $y = x$

c) through the point $(-7, 4)$ and perpendicular to $y = -5$

Practice Problems: *Parallel and Perpendicular Lines*

Complete the table by finding the appropriate slopes.

	Given Line	Slope of Parallel Line	Slope of Perpendicular Line
1.	$y = 4x - 5$		
2.	$x = y + 4$		
3.	$3y = x + 6$		
4.	$x + 2y = 8$		

Write the slope-intercept form of the equation of the line described.

5. line goes through $(3, 4)$ and parallel to $y = \dfrac{9}{2}x - 5$

6. line goes through $(1, -1)$ and parallel to $y = -\dfrac{3}{4}x + 3$

7. line goes through $(-3, -5)$ and perpendicular to $x + 2y = -4$

8. line goes through $(2, -3)$ and perpendicular to $-2x + 5y = -10$

9. line goes through $(1, -5)$ and perpendicular to $-x + y = 1$

10. line goes through $(1, 4)$ and parallel to $y = \dfrac{7}{5}x + 4$

11. line goes through $(-1, 3)$ and parallel to $y = -3x - 1$

12. line goes through $(-4, -1)$ and parallel to $y = -\dfrac{1}{2}x + 1$

13. line goes through $(2, 5)$ and parallel to $x = 0$

14. line goes through $(1, -2)$ and perpendicular to $-x + 2y = 2$

15. line goes through $(5, 2)$ and perpendicular to $5x + y = -3$

16. line goes through $(-8, 3)$ and parallel to $x = 4$

17. line goes through $(1, 3)$ and parallel to $-x + y = 1$

18. line goes through $(7, -3)$ and perpendicular to $x = -2$

19. line goes through $(-2, 5)$ and perpendicular to $y - 2x = 0$

20. line goes through $(-3, -5)$ and perpendicular to $3x + 7y = 0$

 Rescue Roody!

Roody has to find the equation of the line going through the point $(-5, 3)$ and perpendicular to $y = 4$. He is stumped because the line $y = 4$ has a slope of 0 but the line perpendicular to it has an undefined slope. Help him figure out how to work this problem.

2.7. LINEAR INEQUALITIES IN TWO VARIABLES

Objective: Graph the set of solutions to a linear inequality in two variables.

In Chapter 1, we solved linear inequalities in one variable such as $2x - 1 < 3$. The set of solutions, $x < 2$, is the set of all real numbers that satisfy the inequality $2x - 1 < 3$. We can graph this set on the number line.

In this section, we will solve linear inequalities in two variables such as $3x + 4y > 7$ and $y < 2 - 5x$. Since there are two variables, we now need to specify an ordered pair whose coordinates satisfy the inequality.

> **A solution to a linear inequality in two variables is an ordered pair whose coordinates satisfy the inequality after substitution.**

Problem 1 : | *Worked Example*

Are the following ordered pairs solutions to the inequality $y > 2x - 3$?

a) $(2, 3)$ b) $(1, -6)$ c) $(-1, -5)$

   Solution.

In each of these problems, we substitute the coordinates of the given ordered pair into the inequality and determine whether they satisfy the inequality.

a) For the ordered pair $(2, 3)$, we substitute $x = 2$ and $y = 3$ into the inequality

$$y > 2x - 3$$
$$3 > 2(2) - 3$$
$$3 > 4 - 3$$
$$3 > 1 \quad \text{True}$$

Since the coordinates satisfy the inequality, the ordered pair $(2, 3)$ is a solution to the inequality.

b) For the ordered pair $(1, -6)$, we substitute $x = 1$ and $y = -6$ into the inequality $y > 2x - 3$:

$$-6 > 2(1) - 3$$
$$-6 > 2 - 3$$
$$-6 > -1 \quad \text{False}$$

Since the coordinates do not satisfy the inequality, the ordered pair $(0, -6)$ is not a solution to the inequality.

c) For the ordered pair $(-1, -5)$, we substitute $x = -1$ and $y = -5$ into the inequality $y > 2x - 3$:

$$-5 > 2(-1) - 3$$
$$-5 > -2 - 3$$
$$-5 > -5 \quad \text{False}$$

Since the coordinates do not satisfy the inequality, the ordered pair $(-1, -5)$ is not a solution to the inequality.

Problem 2 : | *You Try*

Are the ordered pairs solutions to the inequality $3x + 4y < 12$?

a) $(-1, 5)$ b) $(0, 0)$ c) $\left(-1, \dfrac{15}{4}\right)$

We now wish to describe the set of all ordered pair solutions to a linear inequality in two variables. Let us return to the inequality $y > 2x - 3$ from Worked Example #1. Below is the graph of the line $y = 2x - 3$ and the ordered pairs $(2, 3)$, $(0, -6)$, and $(-2, -7)$.

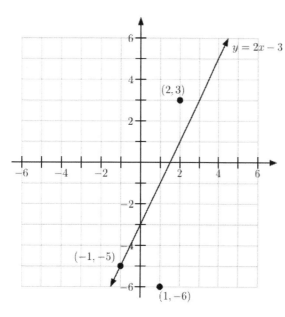

• First, we note that the point $(-1, -5)$ is on the line $y = 2x - 3$. This is because its coordinates satisfy the *equation* $y = 2x - 3$. In fact, all points on the line $y = 2x - 3$ are precisely those whose coordinates satisfy the equation $y = 2x - 3$.

• Next, we see that the point $(2, 3)$ lies above the line $y = 2x - 3$. Because of this, the point $(2, 3)$ satisfies the inequality $y > 2x - 3$ (which we checked in Worked Example #1:

$$3 > 2(2) - 3 \Rightarrow 3 > 1, \text{True}).$$

In fact, any point that lies above the line $y = 2x - 3$ satisfies the inequality $y > 2x - 3$ since its y-value is greater than the corresponding y-value of the point on the line directly below it.

- Finally, we see that the point $(1, -6)$ lies below the line $y = 2x - 3$. Because of this, the point $(1, -6)$ satisfies the inequality $y < 2x - 3$ (which we checked in Worked Example #1: $-6 < 2(1) - 3 \Rightarrow -6 < -1$, True). In fact, any point that lies below the line $y = 2x - 3$ satisfies the inequality $y < 2x - 3$ since its y-value is less than the corresponding y-value of the point on the line directly above it.

We summarize this in the graph below.

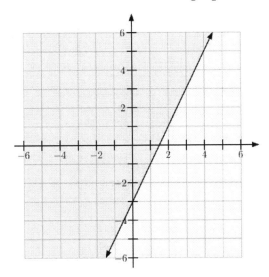

To graph the inequality, $y > 2x - 3$, note that we do not include the points on the line $y = 2x - 3$. Because of this, we will draw the graph of the line $y = 2x - 3$ with a dashed line.

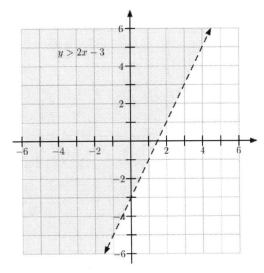

The above observations lead us to the following method for graphing the set of solutions to a linear inequality in two variables.

To graph the set of solutions to a linear inequality in two variables:

- Graph the boundary line by replacing the inequality with an equality. If the inequality is strict ($<$ or $>$), use a dashed line. If the inequality is not strict ($\leq$ or $\geq$), use a solid line.

- Isolate y on the left side of the inequality (if necessary). If the inequality is a "greater than" ($>$) or a "greater than or equal to" ($\geq$), shade above the line. If the inequality is a "less than" ($<$) or a "less than or equal to" ($\leq$), shade below the line.

Problem 3 : | *Worked Example*

Graph the set of solutions to the inequality.

a) $y < -\frac{1}{2}x + 4$ b) $3x - 4y < 12$ c) $y > -2$ d) $x < 4$

   Solution.

a)

We first graph the line $y = -\frac{1}{2}x + 4$.
This is done using the slope $m = -\frac{1}{2}$
and the y-intercept $(0, 4)$. Because the inequality is non-strict ($\leq$), we include the points on the line, and so we use a solid line.
Also, note that we already have y isolated on the left side of the inequality. Then, since the inequality is a "less than or equal to" sign, we shade the region below the line.

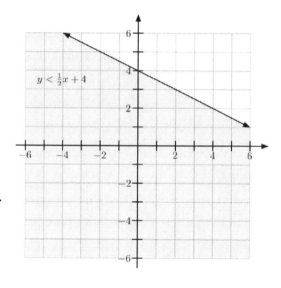

b) We first graph the line $3x - 4y = 12$. This is easily done using the intercepts, $(4, 0)$ and $(0, -3)$. Because the inequality is strict $(<)$, we do not include the points on the line, and so use a dashed line. Unlike part (a), we do not have y isolated on the left side of the inequality. We do this below:

$$3x - 4y < 12$$
$$-4y < -3x + 12$$
$$y > \frac{3}{4}x - 3$$

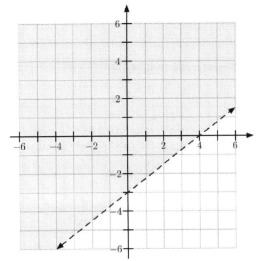

Then, since the inequality is a "greater than" $(>)$, we shade the region above the line.

c) We first graph the line $y = -2$, which is a horizontal line. Because the inequality is non-strict $(\geq)$, we use a solid line. Also, note that y is isolated on the left side of the inequality. Then, since the inequality is a "greater than or equal to" $(\geq)$, we shade the region above the line.

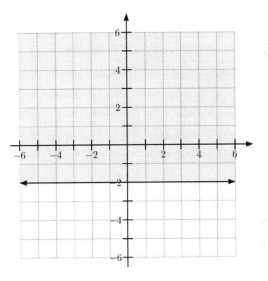

d) We first graph the line $x = 4$, which is a vertical line. Because the inequality is strict $(<)$, we use a dashed line. Note that in this example, there isn't a y-term present, and so we cannot isolate y on the left side of the inequality. However, reading the inequality $x < 4$ as "x is less than 4" indicates that we should shade the region to the left of the line.

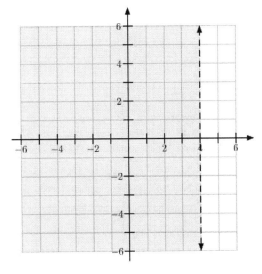

Problem 4 : | *Media/Class Example*

Graph the set of solutions to the inequality.

a) $4x + 3y < 7$

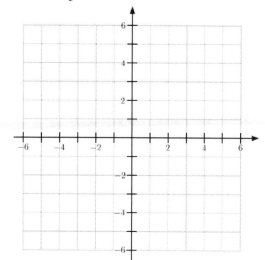

c) $y - 3 > 0$

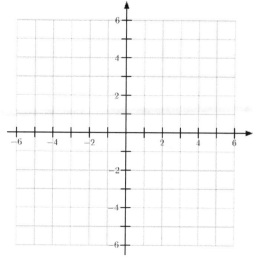

b) $x - y < 2$

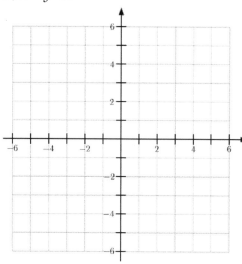

d) $x + 4 > 0$

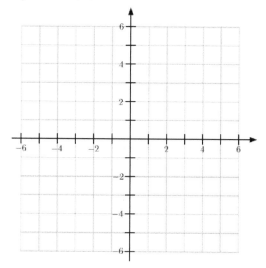

Problem 5 : | You Try

Graph the set of solutions to the inequality.

a) $y < 2 + 3x$

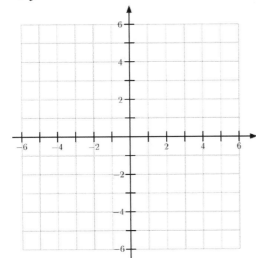

c) $x > -2$

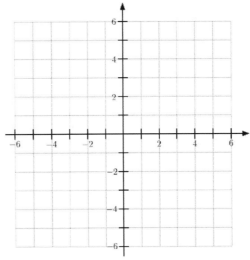

b) $x > y$

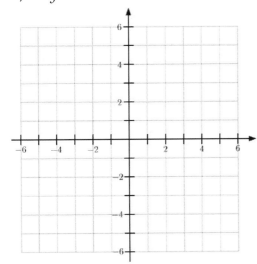

d) $2x + 5y < 7$

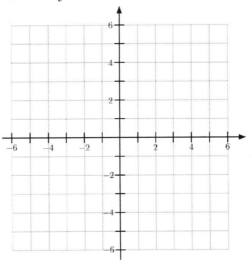

Practice Problems: *Linear Inequalities in Two Variables*

Graph the set of solutions to the inequality.

1. $y \geq -\dfrac{1}{3}x - 2$

2. $y \leq \dfrac{2}{3}x$

3. $y < 3x - 1$

4. $x < y$

5. $y \geq -2$

6. $3 < x$

7. $y + 4x \leq 5$

8. $x - y \leq -1$

9. $3x + y \geq 0$

10. $2y < \dfrac{1}{2}x$

11. $-x + 3y \geq 6$

12. $x - 2y > -4$

13. $y - 4 < 0$

14. $x + 1 < -1$

CHAPTER 2 ASSESSMENT

1. Given the line $y = -\dfrac{4}{5}x + 4$.

    a) Identify the slope of the line.

    b) Identify the x-intercept. Write your answer as ordered pairs.

    c) Identify the y-intercept. Wrtie your answer as ordered pairs.

2. Find the slope of the line through each pair of points.

    a) $(0, -3)$ and $(4, 3)$

    b) $(-7, -7)$ and $(9, -6)$

3. Write the equation of the line with the following properties.

    a) line has slope $-\dfrac{5}{2}$ and y-intercept $(0, 6)$

    b) line goes through $(6, 3)$ and $(4, -2)$

    c) horizontal line going through $(5, -8)$

    d) line goes through $(9, -3)$ and is parallel to $2x + y = 1$

    e) line goes through $(0, -7)$ and is perpendicular to $y = \dfrac{2}{3}x + 5$

4. Sketch the graph of each line.

    a) $y = \dfrac{1}{2}x - 1$

    b) $3x + y = 0$

5. Sketch the graph of each inequality.

    a) $y > \dfrac{2}{3}x - 4$

    b) $x - 2y \geq 2$

CHAPTER 3

SYSTEMS OF LINEAR EQUATIONS

3.1. INTRODUCTION TO SYSTEMS OF LINEAR EQUATIONS

Objective: To determine if a given point is a solution of a system of equations.

| Example 1 | *Worked Example* |

A car-stereo warehouse has two options of compensation for their employees:

- Plan A pays $200 a week plus a $30 commision per stereo sold.
- Plan B has no salary but pays a $50 commission per stereo sold.

Paul is employed under plan A and Bob is employed under plan B. Last week Paul and Bob were both paid the same amount (before deductions). How many stereos did each one of them sell last week?

Solution.

Notice that each plan can be represented by a linear equation as

$$y = 30x + 200 \quad \text{for plan A}$$
$$y = 50x \quad\quad\quad \text{for plan B}$$

where x is the number of stereos sold last week, and y is the total salary paid last week.

The problem is asking to find the value of x that will produce the same y-value for both equations. Let's explore some values for each equation:

Plan A

x	y
1	230
2	260
3	290
5	350
10	500

Plan B

x	y
1	50
2	100
3	150
5	250
10	500

We found our solution: each one sold 10 car stereos last week.◇

Notice that the point $(10, 500)$ is a solution for **both equations**:

Plan A

$$500 \overset{?}{=} 30(10) + 200$$
$$500 = 300 + 200 \quad \checkmark$$

Plan B

$$y \overset{?}{=} 50(10)$$
$$y = 500 \quad \checkmark$$

In situations as above we will call the set of two equations

$$y = 30x + 200$$
$$y = 50x$$

a **system of two linear equations**. The point $(10, 500)$ **is a solution** to this system. In this chapter we will learn more about systems of two linear equations, their solutions, and how to find them. The method presented above is not the best way to find a solution, and will not work for two general linear equations.

Notice that since each equation represents a line in the xy-axis, a solution to a system of two linear equations, if it exists, will be exactly one point. It could also happen that the two lines do not have a common point, in which case we will say that the system has **no solutions**.

Example 2 | *Worked Example*

Determine if the point $(-2, 6)$ is a solution to the system of equations

$$3x + 2y = 6$$

$$x - y = -8$$

Solution.

In order to be a solution to the system, the point $(-2, 6)$ must be a solution to each line. Let's check

$$3(-2) + 2(6) \overset{?}{=} 6 \qquad -2 - 6 \overset{?}{=} -8$$
$$-6 + 12 \overset{?}{=} 6 \checkmark \qquad -8 = -8 \checkmark$$

So indeed, the point $(-2, 6)$ is the solution to the system.

Example 3 | *Worked Example*

Determine if the point $(-2, -4)$ is the solution to the system of equations

$$5x - 4y = 6$$

$$5x - 2y = 2$$

Solution.

In order to be a solution to the system, the point $(-2, -4)$ must be a solution to each line. Let's check

$$5(-2) - 4(-4) \overset{?}{=} 6 \qquad 5(-2) - 2(-4) \overset{?}{=} 2$$
$$-10 + 16 \overset{?}{=} 6 \checkmark \qquad -10 + 8 \neq -2$$

So the point $(-2, -4)$ is a not the solution to the system.

Note. In the last example we do not claim that the system does not have a solution, in fact it does. We are only saying that the point $(-2, 4)$ is not the solution. In the following sections we will learn methods to find the solution for this kind of problems.

Example 4	*Class/Media Example*

Determine if the point $(-5, 13)$ is a solution to the system of equations

$$4x + 2y = 6$$

$$x + y = 8$$

Example 5	*You Try*

Determine if the point $(5, 7)$ is a solution to the system of equations

$$-x + y = 2$$

$$-6x + 10y = 40$$

Example 6	*Worked Example*

Determine if the point $\left(\dfrac{31}{10}, \dfrac{3}{5}\right)$ is a solution to the system of equations

$$2x - 2y = 5$$

$$2x + 3y = 8$$

Solution.

Let's check if the point $\left(\dfrac{31}{10}, \dfrac{3}{5}\right)$ is a solution to each equation.

$$2\left(\frac{31}{10}\right) - 2\left(\frac{3}{5}\right) \overset{?}{=} 5 \qquad\qquad 2\left(\frac{31}{10}\right) + 3\left(\frac{3}{5}\right) \overset{?}{=} 8$$

$$\frac{31}{5} - \frac{6}{5} \overset{?}{=} 5 \qquad\qquad\qquad \frac{31}{5} + \frac{9}{5} \overset{?}{=} 8$$

$$\frac{25}{5} = 5 \checkmark \qquad\qquad\qquad \frac{40}{5} \overset{?}{=} 8 \checkmark$$

Example 7 | *Class/Media Example*

Determine if the point $\left(\dfrac{-7}{3}, \dfrac{5}{3}\right)$ is a solution to the system of equations

$$-4x - 2y = 6$$

$$-x + y = 4$$

Example 8 | *You Try*

Determine if the point $\left(\dfrac{1}{4}, -\dfrac{9}{4}\right)$ is a solution to the system of equations

$$5x - 3y = 8$$

$$-x - y = 2$$

Practice Problems: *Introduction to Systems of Linear Equations*

Determine if the given point is a solution to the corresponding Systems of Equations

1. $(2,1)$;
$$3x - y = 5$$
$$2x + 3y = 7$$

2. $(2,-8)$;
$$y = -4x$$
$$x - y = 10$$

3. $(2,-2)$;
$$3x - 2y = 10$$
$$4x + 5y = 8$$

4. $(6,-2)$;
$$x + y = 4$$
$$x - 3y = 12$$

5. $(1145,1345)$;
$$x - y = -200$$
$$13x - 11y = 90$$

6. $\left(-19, -\dfrac{85}{2}\right)$;
$$x - 2y = 66$$
$$3x - 2y = 28$$

7. $(2,-2)$;
$$3x - 2y = 10$$
$$4x + 5y = 8$$

8. $(10,10)$;
$$x = 2y - 10$$
$$y = -15 + 3x$$

9. $\left(\dfrac{-1}{2}, \dfrac{-5}{4}\right)$;
$$-x - 2y = 3$$
$$x - 2y = 2$$

10. $(4,-5)$;
$$4x - 5y = 41$$
$$-3x + 2y = 18$$

3.2. SOLVING SYSTEMS OF LINEAR EQUATIONS BY GRAPHING

Objective: Solve systems of linear equations by graphing.

In this section we will find the solutions to systems of two linear equations by graphing both lines and finding their common point on the graph. This method will not always work, but it is a good starting point to grasp a good understanding of the material.

Example 1 *Worked Example*

Find the point where the following two lines intersect: $\begin{cases} 3x - y = 5 \\ x + y = 3 \end{cases}$

Solution: We will graph each line and in order to do this we will re-write the lines in slope-intercept form.
Our first line becomes

$$
\begin{aligned}
3x - y &= 5 \\
\underline{-3x \quad -3x} & \qquad \text{subtract } 3x \text{ from each side} \\
-y &= -3x + 5 \\
(-1)(-y) &= (-1)(-3x + 5) \quad \text{multiply by } (-1) \text{ each side} \\
y &= 3x - 5
\end{aligned}
$$

So our first line has slope $m = 3$, and y-intercept at $(0, -5)$.
Our second line becomes

$$
\begin{aligned}
x + y &= 3 \\
\underline{-x \quad = \quad -x} & \qquad \text{subtract } x \text{ from each side} \\
y &= -x + 3
\end{aligned}
$$

So our second line has slope $m = -1$ and y-intercept at $(0, 3)$.
We can now graph both lines on the same plane.

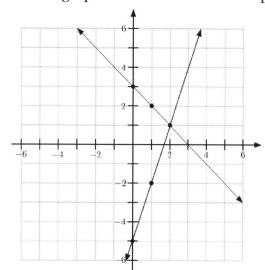

We can see from the graph that

-The points $(0, -5)$ and $(2, 1)$ create a slope of 3, so the point $(2, 1)$ belongs to the first line.

-The points $(0, 3)$ and $(2, 1)$ create a slope of -1, so the point $(2, 1)$ belongs to the second line as well.

Since the point $(2,1)$ belongs to both lines, it is a solution to each equation. We check this below :

1st equation	2nd equation
$3x - y = 5$	$x + y = 3$
$3(2) - (1) = 5$	$(2) + (1) = 3$

The point $(2,1)$ is the point where these two lines intersect.

In the above example the *system of equations* was the set of the two original equations

$$3x - y = 5$$
$$x + y = 3$$

and *the solution* is the pair $(x, y) = (2, 1)$.

Problem 1 : | *Class/Media Example*

Solve the system of equations by graphing: $6x - 3y = -9$
$2x + 2y = -6$

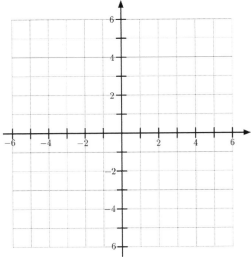

Problem 2 : | *You Try*

Solve the system of equations by graphing: $\begin{array}{l} x+y=0 \\ x-2y=-6 \end{array}$

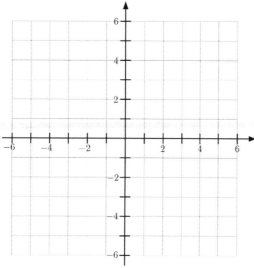

Problem 3 : | *You Try*

Solve the system of equations by graphing: $\begin{array}{l} 2x-y=2 \\ x-2y=-2 \end{array}$

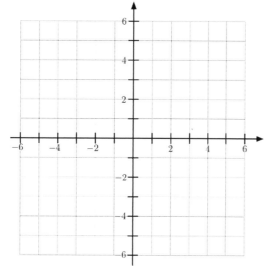

Problem 4 : | *Worked Example*

Solve the system of equations by graphing: $\begin{aligned} 2x-3y &= 12 \\ 4x-6y &= -6 \end{aligned}$

Solution Let's first re-write the equations in slope-intercept form:

$$
\begin{aligned}
2x-3y &= 12 \\
-3y &= -2x+12 \\
y &= \frac{-2x+12}{(-3)} \\
y &= \frac{2}{3}x-4
\end{aligned}
\qquad\qquad
\begin{aligned}
4x-6y &= -6 \\
-6y &= -4x-6 \\
y &= \frac{-4x-6}{-6} \\
y &= \frac{2}{3}x+1
\end{aligned}
$$

Now let's look at the graphs:

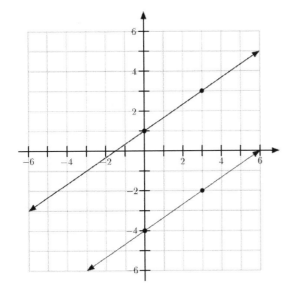

Both lines have the same slope, $m=\dfrac{2}{3}$, and have different y-intercepts, so the lines will not intersect.

In this case we will say that there is **no solution** to the system of equations.

The symbol $\emptyset$ is used in mathematics to denote the *empty set*, the set that has no elements. Since the solution to the above system has *no elements*, we can refer to the solution to the above system of equations as being empty, or $\emptyset$.

Problem 5 : | *Worked Example*

Solve the system of equations by graphing: $\begin{array}{l} 4x + 3y = -12 \\ -8x - 6y = 24 \end{array}$

Solution Let's first re-write the equations in slope-intercept form:

$$4x + 3y = -12$$
$$3y = -4x - 12$$
$$y = \frac{-4x - 12}{3}$$
$$y = \frac{-4}{3}x - 4$$

$$-8x - 6y = 24$$
$$-6y = 8x + 24$$
$$y = \frac{8x + 24}{-6}$$
$$y = \frac{8}{(-6)}x - 4$$
$$y = \frac{-4}{3}x - 4$$

Both equations represent the same line!

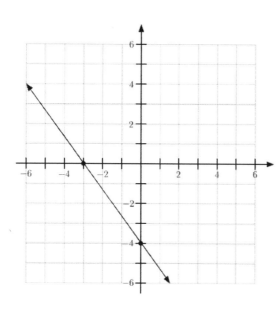

Any point on the first line is also on the second line, that is why we see only one line graphed.

In this case we will say that there are **infinitely many solutions** to the system of equations.

Problem 6 : | *Class/Media Example*

Solve the system of equations by graphing: $\begin{aligned} 2x+5y &= -10 \\ 8x+20y &= -40 \end{aligned}$

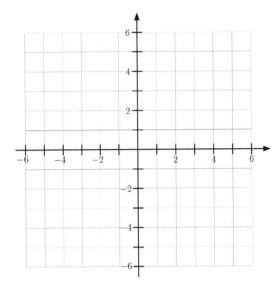

Problem 7 : | *You Try*

Solve the system of equations by graphing: $\begin{aligned} x+3y &= 15 \\ 2x+6y &= 24 \end{aligned}$

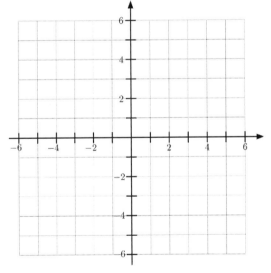

Problem 8 : | *You Try*

Solve the system of equations by graphing: $\begin{array}{l} 2x + 3y = 9 \\ -6x = 9y - 27 \end{array}$

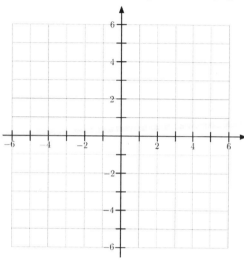

Problem 9 : | *You Try*

Solve the system of equations by graphing: $\begin{array}{l} 3x - 2y = -2 \\ x + 2y = 6 \end{array}$

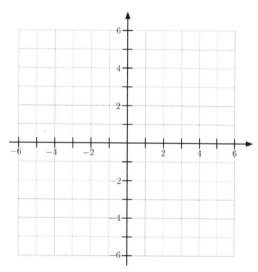

As you can see from the last example, this method will not be the best way to solve a system of equations. In general, two lines can intersect at a point that will not be easy to find from a graph. In the next sections we will learn other methods to solve a system that do not depend on the graphs of the lines.

World View Note: The Babylonians were the first to work with systems of equations with two variables. However, their work with systems was quickly passed by the Greeks who would solve systems of equations with three or four variables and around 300 AD, developed methods for solving systems with any number of unknowns!

Practice Problems: *Solving Systems of Equations by Graphing*

Solve the following systems of linear equations by graphing.

1. $\begin{aligned} -x+y &= 1 \\ -3x+y &= -1 \end{aligned}$

2. $\begin{aligned} -x+2y &= 8 \\ 2x+y &= -1 \end{aligned}$

3. $\begin{aligned} y &= \frac{x}{2}+1 \\ x-y &= -3 \end{aligned}$

4. $\begin{aligned} -x+y &= 1 \\ 2x-2y &= 3 \end{aligned}$

5. $\begin{aligned} x+3y &= 0 \\ x-y &= 4 \end{aligned}$

6. $\begin{aligned} 5x+3y &= 6 \\ 2-y &= \frac{5}{3}x \end{aligned}$

7. $\begin{aligned} 3x-y &= 4 \\ x-2y &= -2 \end{aligned}$

8. $\begin{aligned} y &= \frac{2}{3}x-4 \\ x+y &= 1 \end{aligned}$

9. $\begin{aligned} 4y &= 3x-12 \\ 2x+8y &= 8 \end{aligned}$

10. $\begin{aligned} x-y &= 3 \\ y &= -2 \end{aligned}$

3.3. SUBSTITUTION METHOD

Objective: Solve systems of linear equations using substitution.

Solving a system by graphing has limitations. First, it requires the graph to be perfectly drawn. If the lines are not straight we may arrive at the wrong answer. Second, graphing is not a great method to use if the answer contains a decimal or fraction. For these reasons, we need an algebraic approach.

In this section, the algebraic approach we will learn is **substitution**. The strategy is:

- Solve one of the equations for one of the variables (we call it the "substitution equation").

- Substitute the expression for that variable into the *other equation*. This should produce an equation in one variable.

- Solve for that variable.

- Substitute the known value into the "substitution equation" and solve for the other variable.

- Check your solutions in the original equations.

Problem 1 : | *Worked Example*

Given the equation $y = 2x - 3$, solve for y when $x = 5$.

Solution.

$$\begin{aligned} y &= 2x - 3 & &\text{Substitute } x \text{ with } 5 \\ y &= 2(5) - 3 & &\text{Multiply first} \\ y &= 10 - 3 & &\text{Subtract} \\ y &= 7 & &\text{Our Solution} \end{aligned}$$

When we know the value of one variable, we can substitute that value (or expression) in for the variable in the other equation. It is very important to use parenthesis when we substitute.

Problem 2 : | *Worked Example*

Solve the following system of linear equations: $\begin{array}{l} 2x-3y=7 \\ y=3x-7 \end{array}$

Solution.

We know $y=3x-7$, so we can substitute this into the other equation.

$$2x-3(3x-7)=7 \quad \text{Distribute} -3$$
$$2x-9x+21=7 \quad \text{Combine like terms}$$
$$-7x+21=7 \quad \text{Subtract } 21$$
$$-7x=-14 \quad \text{Divide by} -7$$
$$x=2 \quad \text{We have our } x!$$

$$y=3(2)-7 \quad \text{Substitute } x=2 \text{ into } y= \text{equation to find } y$$
$$y=6-7 \quad \text{Simplify}$$
$$y=-1 \quad \text{We have our } y!$$
$$(2,-1) \quad \text{Our Solution}$$

Verify that we have the correct solution by substituting $(x, y) = (2, -1)$ into each of the original equations.

1st Equation	2nd Equation
$2(2)-3(-1) \overset{?}{=} 7$	$-1 \overset{?}{=} 3(2)-7$
$4+3 = 7\checkmark$	$-1 = 6-7\checkmark$

Problem 3 : | *Media/Class Example*

Solve the following system of linear equations: $\begin{cases} y=x \\ x+y=6 \end{cases}$

Problem 4 : | *You Try*

Solve the following system of linear equations. Be sure to check your answer.

a) $\begin{cases} y = 2x + 5 \\ y - 4x = 4 \end{cases}$

b) $\begin{cases} y = x + 3 \\ x - \dfrac{1}{2}y = -2 \end{cases}$

c) $\begin{cases} 3x - 4y = 9 \\ x = y + 3 \end{cases}$

Sometimes there is no isolated variable in either equation, so we need to first isolate a variable. It does not matter which variable or which equation you choose. However, once you choose an equation, you want to make sure you substitute into the *other equation*.

Problem 5 : | Worked Example

Solve the following system of linear equations: $\begin{cases} 3x+2y=1 \\ x-5y=6 \end{cases}$

Solution.

The second equation, $x-5y=6$ has a lone variable, x. Solve for x by subtracting $5y$ from each side to get $x=6+5y$. Substitute this into the other equation.

$$3(6+5y)+2y=1 \quad \text{Substitute } x=6+5y \text{ and distribute}$$
$$18+15y+2y=1 \quad \text{Combine like terms}$$
$$18+17y=1 \quad \text{Subtract 18 from each side}$$
$$17y=-17 \quad \text{Divide each side by 17}$$
$$y=-1 \quad \text{We have our } y!$$

$$x=6+5(-1) \quad \text{Substitute } y=-1 \text{ into } x=\text{equation to find } x$$
$$x=6-5 \quad \text{Simplify}$$
$$x=1 \quad \text{We have our } x!$$
$$(1,-1) \quad \text{Our Solution}$$

Verify that we have the correct solution by substituting $(x, y) = (1, -1)$ into each of the original equations.

First Equation

$$3(1)+2(-1) \overset{?}{=} 1$$
$$3-2 = 1\checkmark$$

Second Equation

$$(1)-5(-1) \overset{?}{=} 6$$
$$1+5 = 6\checkmark$$

Problem 6 : | Media/Class Example

Solve the following system of linear equations: $\begin{cases} 3x-2y=2 \\ x+4y=3 \end{cases}$

Problem 7 : | *You Try*

Solve the following system of linear equations. Be sure to check your answer.

a) $\begin{cases} x+y=5 \\ x-y=-1 \end{cases}$
b) $\begin{cases} \frac{1}{2}x+3y=-2 \\ x+2y=4 \end{cases}$

Sometimes it takes more than one step to isolate a variable. Choose your variables wisely.

Problem 8 : | *Worked Example*

Solve the following system of linear equations: $\begin{cases} 5x-6y=-14 \\ -2x+4y=12 \end{cases}$

Solution.

None of the equations have a lone variable. We will solve for x in the second equation because the coefficient of x and y, including the constant term are even.

$$-2x+4y=12 \quad \text{Divide every term by } -2$$
$$x-2y=-6 \quad \text{Add } 2y \text{ to each side to solve for } x$$
$$x=2y-6$$

Substitute $x=2y-6$ into the first equation and solve for y.

$$5(2y-6)-6y=-14 \quad \text{Substitute } x=2y-6 \text{ and distribute}$$
$$10y-30-6y=-14 \quad \text{Combine like terms}$$
$$4y-30=-14 \quad \text{Add } 30 \text{ to each side}$$
$$4y=16 \quad \text{Divide each side by } 4$$
$$y=4 \quad \text{We have our } y!$$

$$x=2(4)-6 \quad \text{Substitute } y=4 \text{ into } x=\text{equation to find } x$$
$$x=8-6 \quad \text{Simplify}$$
$$x=2 \quad \text{We now have our } x!$$
$$(2,4) \quad \text{Our Solution}$$

Verify that we have the correct solution by substituting $(x, y) = (2, 4)$ into each of the original equations.

First Equation

$$5(2) - 6(4) \overset{?}{=} -14$$
$$10 - 24 = -14\checkmark$$

Second Equation

$$-2(2) + 4(4) \overset{?}{=} 12$$
$$-4 + 16 = 12\checkmark$$

Problem 9 : | *Media/Class Example*

Solve the following system of linear equations: $\begin{cases} 2x - 4y = 8 \\ 9x + 3y = -6 \end{cases}$

Problem 10 : | *You Try*

Solve the following system of linear equations: $\begin{cases} 2y - 8x = 6 \\ 2x + 3y = 2 \end{cases}$

As we saw with graphing, it is possible that a system has no solution, $\emptyset$ (parallel lines) or infinitely many solution (same line). When solving these kinds of systems algebraically, the process takes an interesting turn.

Problem 11 : | *Worked Example*

Solve the following system of linear equations: $\begin{cases} y = 3x - 4 \\ 2y - 6x = -8 \end{cases}$

    Solution.

Substitute the first equation, $y = 3x - 4$ into the second equation.

$$2(3x - 4) - 6x = -8 \quad \text{Distribute}$$
$$6x - 8 - 6x = -8 \quad \text{Combine like terms}$$
$$-8 = -8 \quad \text{Variables are gone! True statement.}$$
$$\text{Infinitely many solutions} \quad \text{Our Solution}$$

Because we had a true statement, and no variables, we know that anything that works in the first equation, will also work in the second equation. However, we do not always end up with a true statement.

Problem 12 : | *Worked Example*

Solve the following system of linear equations: $\begin{cases} 6x - 3y = -9 \\ y = 2x + 5 \end{cases}$

    Solution.

Substitute the second equation, $y = 2x + 5$ into the first equation.

$$6x - 3(2x + 5) = -9 \quad \text{Dstribute}$$
$$6x - 6x - 15 = -9 \quad \text{Combine like terms}$$
$$-15 \neq -9 \quad \text{Variables are gone! False statement.}$$
$$\text{No Solution or } \emptyset \quad \text{Our Solution}$$

Because we had a false statement, and no variables, we know that nothing will work in both equations.

Problem 13 : | *You Try*

Solve the following system of linear equations.

a) $\begin{cases} y = 4 - 2x \\ 6x + 3y = 12 \end{cases}$
b) $\begin{cases} x = -y \\ y = 4 - x \end{cases}$

Practice Problems: *Substitution Method*

Solve each system of linear equations by substitution. Be sure to check your answers.

1. $y = -3x$
   $y = 6x - 9$

2. $b = a + 5$
   $b = -2a - 4$

3. $q = -6p + 3$
   $q = 6p + 3$

4. $3h - g = 12$
   $g = 6h + 21$

5. $6m - 4n = -8$
   $n = -6m + 2$

6. $-2x + 2y = 18$
   $y = 7x + 15$

7. $m - n = -4$
   $3m - 4n = -19$

8. $d = -8c + 19$
   $-c + d = 16$

9. $2p + q = 8$
   $-7p - 6q = -8$

10. $a - b = 3$
    $-3a + 3b = 6$

11. $a - 5b = 7$
    $2a + 7b = -20$

12. $-2c - d = -5$
    $c - d = -23$

13. $6x + 4y = 16$
    $-4x + 2y = -6$

14. $-6x + y = 20$
    $18x - 3y = -18$

15. $7x + 5y = -13$
    $\dfrac{1}{4}x - y = -4$

16. $2a + b = 5$
    $4a - 5b = -4$

17. $x - 2y = -2$
    $y = \dfrac{1}{2}x + 1$

18. $x + 5y = 15$
    $-3x + 2y = 6$

19. $2m + 3n = -10$
    $7m + n = 3$

20. $2x + 6y = 2$
    $4y - 2y = -3$

21. $-6m + 6n = -12$
    $8m - 3n = 16$

22. $2a + 3b = 16$
    $-7a - b = 20$

For each of the graphs below, do the following:

a) Find the solution from the given graphs.

b) Find the equation of each line.

c) Using substitution, solve the system found in (b) to verify the solution found in (a).

23.

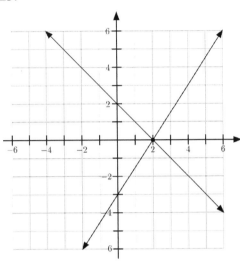

25.

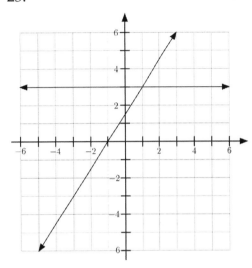

24.

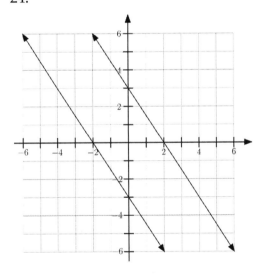

26.

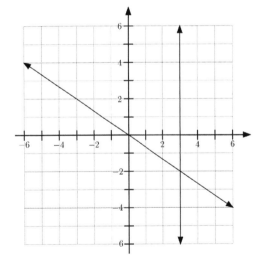

Rescue Roody!

Roody is told to solve the following system of linear equations: $\begin{cases} y = x - 3 \\ 2x - y = 5 \end{cases}$. Roody substitutes $y = x - 3$ into the second equation, as follows:

$$2x - x - 3 = 5$$
$$x - 3 = 5$$
$$x = 8$$

Roody substitutes $x = 8$ into $y = x - 3$:

$$y = 8 - 3$$
$$y = 5$$

Roody's answer is $(x, y) = (8, 5)$. He checks his answer:

$$y = x - 3$$
$$5 = 8 - 3 \checkmark$$

Roody compares his answer with a classmate but their answers do not match. What happened?

3.4. ELIMINATION METHOD

Objective: Solve systems of linear equations using the elimination method.

The third process used to solve a system of linear equations is the elimination method. This process is especially useful when the system of linear equations is written in the form $Ax+By=C$. If we can eliminate one of the two variables, we can readily solve the equation.

Problem 1 : | Worked Example

Solve the following system of linear equations: $\begin{cases} 3x-4y=8 \\ 5x+4y=-24 \end{cases}$

Solution.

Notice the coefficients of the y terms are opposites.

$$\begin{cases} 3x-4y=8 \\ \underline{5x+4y=-24} \end{cases} \quad \text{Add columns}$$

$$8x=-16 \quad \text{Solve for } x, \text{ divide by } 8$$

$$x=-2 \quad \text{We have our } x!$$

Substitute $x=-2$ into one of the original equations to solve for y.

$$5(-2)+4y=-24 \quad \text{Simplify}$$

$$-10+4y=-24 \quad \text{Add 10 to each side}$$

$$4y=-14 \quad \text{Divide by 4}$$

$$y=-\frac{7}{2} \quad \text{We have our } y!$$

$$\left(-2,-\frac{7}{2}\right) \quad \text{Our Solution}$$

Verify that we have the correct solution by substituting $(x,y)=\left(-2,-\frac{7}{2}\right)$ into each of the original equations.

First Equation

$$3(-2)-4\left(-\frac{7}{2}\right) \overset{?}{=} 8$$

$$-6+14 = 8\checkmark$$

Second Equation

$$5(-2)+4\left(-\frac{7}{2}\right) \overset{?}{=} -24$$

$$-10-14 = -24\checkmark$$

Problem 2 : | *Media/Class Example*

Solve the following system of linear equations: $\begin{cases} -7a+b=-10 \\ -9a-b=-22 \end{cases}$

Problem 3 : | *You Try*

Solve the following system of linear equations: $\begin{cases} m-n=4 \\ 2m+n=8 \end{cases}$ and be sure to check your answers.

With the previous example, the y terms eliminated readily because the coefficients were opposites, and the sum of opposites equal zero. We now examine a system in which the coefficients are not opposites. If this is the case, we re-balance our equations so that one of the two variables will eliminate and then follow the process demonstrated in the previous example.

Problem 4 : | *Worked Example*

Solve the following system of linear equations: $\begin{cases} -6x+5y=22 \\ x+\dfrac{3}{2}y=1 \end{cases}$

Solution.

We choose to eliminate x. Remember that you want the coefficient of the variable to be opposites. Multiply the second equation by 6. That way, one coefficient of x is -6 and the other 6, so they are opposites. Note that when multiplying the second equation by 6, you must distribute the 6 to the entire equation.

$$\begin{cases} -6x+5y=22 \\ 6\left(x+\dfrac{3}{2}y\right)=6(1) \end{cases} \quad \text{Multiply the second equation by 6}$$

$$\begin{cases} -6x+5y=22 \\ \underline{6x+9y=6} \\ 14y=28 \end{cases} \quad \text{Add columns}$$

$$14y=28 \quad \text{Divide each side by 14}$$
$$y=2 \quad \text{We have our } y!$$

Substitute $y=2$ into one of the original equations to solve for x.

$$x+\frac{3}{2}(2)=1 \quad \text{Simplify}$$
$$x+3=1 \quad \text{Subtract 3 from each side}$$
$$x=-2 \quad \text{We also have our } x!$$
$$(-2,2) \quad \text{Our Solution}$$

Verify that we have the correct solution by substituting $(x,y)=(-2,2)$ into each of the original equations.

First Equation

$$-6(-2)+5(2) \overset{?}{=} 22$$
$$12+10 = 22\checkmark$$

Second Equation

$$(-2)+\frac{3}{2}(2) \overset{?}{=} 1$$
$$-2+3 = 1\checkmark$$

Problem 5 : | *Media/Class Example*

Solve the following system of linear equations: $\begin{cases} -6m - 2n = -9 \\ -\dfrac{3}{2}m + n = 3 \end{cases}$

Problem 6 : | *You Try*

Solve the following system of linear equations: $\begin{cases} \dfrac{4}{3}x - y = -2 \\ -4x + 5y = 20 \end{cases}$ and be sure to check your answer.

When we consider what number to use so that the equations have one set of variables that are opposites, we generally choose the Least Common Multiple (LCM) of the coefficients of the variable we want to eliminate.

Problem 7 : | *Worked Example*

Solve the following system of linear equations: $\begin{cases} -4x + 5y = 12 \\ -5x + 3y = 15 \end{cases}$

   Solution.

First choose which variable to eliminate. We choose to eliminate y. The LCM of the coefficient of y, 5 and 3, is 15. Multiply the first eqution by 3 and the second equation by -5. That way, one coefficient of y is 15 and the other -15, so they are opposites.

$$\begin{cases} 3(-4x + 5y) = 3(12) & \text{Multiply the first equation by 3} \\ -5(-5x + 3y) = -5(15) & \text{Multiply the second equation by } -5 \end{cases}$$

$$\begin{cases} -12x + 15y = 36 \\ \underline{25x - 15y = -75} \qquad \text{Add columns} \\ 13x = -39 \end{cases}$$

$$13x = -39 \quad \text{Divide each side by 13}$$
$$x = -3 \quad \text{We have our } x!$$

Substitute $x = -3$ into one of the original equations to solve for y.

$$-4(-3) + 5y = 12$$
$$12 + 5y = 12 \quad \text{Subtract 12 from each side}$$
$$5y = 0 \quad \text{Divide each side by 5}$$
$$y = 0 \quad \text{We have our } y!$$
$$(-3, 0) \quad \text{Our Solution}$$

Verify that we have the correct solution by substituting $(x, y) = (-3, 0)$ into each of the original equations.

First Equation $\qquad\qquad\qquad\qquad\qquad\qquad$ Second Equation

$$-4(-3) + 5(0) \overset{?}{=} 12 \qquad\qquad -5(-3) + 3(0) \overset{?}{=} 15$$
$$12 + 0 = 12\checkmark \qquad\qquad\qquad 15 + 0 = 15\checkmark$$

Problem 8 : | *Media/Class Example*

Solve the following system of linear equations: $\begin{cases} \frac{1}{2}g - \frac{1}{4}h = \frac{9}{4} \\ 3g - 8h = 7 \end{cases}$

Problem 9 : | *You Try*

Solve the following system of linear equations: $\begin{cases} 5g - 4h = 1 \\ 11g + 6h = 17 \end{cases}$ and be sure to check your answer.

It is important that the two equations are written in the same form before we begin the process of elimination.

| **Problem 10 :** | *Media/Class Example* |

Solve the following system of linear equations: $\begin{cases} 2x - 5y = -13 \\ -3y + 4 = -5x \end{cases}$

Just as with graphing and substution, it is possible with elimination to have no solution or infinitely many solutions. If all the variables disappear from our problem, a true statment will indicate infinitely many solutions and a false statment will indicate no solution.

| **Problem 11 :** | *Worked Example* |

Solve the following system of linear equations: $\begin{cases} 2x = 5y + 3 \\ -6x + 15y = -9 \end{cases}$

> **Solution.**

The two equations are not written in the same form.

$$\begin{cases} 2x = 5y + 3 \\ -6x + 15y = -9 \end{cases} \qquad \text{Subtract } 5y \text{ from each side of first equation}$$

$$\begin{cases} 2x - 5y = 3 \\ -6x + 15y = -9 \end{cases} \qquad \text{Equations are of the same form}$$

We choose to eliminate x. LCM of the coefficients of x is 6.

$$\begin{cases} 3(2x - 5y) = 3(3) \\ -6x + 15y = -9 \end{cases} \qquad \text{Multiply the first equation by 3}$$

$$\begin{cases} 6x - 15y = 9 \\ \underline{-6x + 15y = -9} \end{cases}$$
$$\qquad\qquad\qquad\quad \text{Add columns}$$
$$0 = 0 \qquad \text{True statement}$$

$$\text{Infinitely many solutions} \qquad \text{Our Solution}$$

Problem 12 : | *You Try*

Solve the following system of linear equations: $\begin{cases} \dfrac{1}{3}x - \dfrac{1}{2}y = \dfrac{2}{3} \\ 0.6x + 0.9y = 1.5 \end{cases}$

Problem 13 : | *You Try*

Solve the following system of linear equations: $\begin{cases} y - x = 3 \\ 2x - 2y = -6 \end{cases}$

We have covered three different methods that can be used to solve a system of two equations with two variables. While all three can be used to solve any system, each method has its own strengths. It is important you are familiar with all three methods.

Practice Problems: *Elimination Method*

Solve each system by elimination.

1. $\begin{cases} 9x + y = 22 \\ 7x - y = 10 \end{cases}$

2. $\begin{cases} 4m + 9n = -28 \\ -4m - 2n = 0 \end{cases}$

3. $\begin{cases} 4a - 6b = -10 \\ 4a - 6b = -14 \end{cases}$

4. $\begin{cases} 2c - d = 5 \\ 5c + 2d = -28 \end{cases}$

5. $\begin{cases} u - 2v = 5 \\ 5u - 6v = 17 \end{cases}$

6. $\begin{cases} x + 3y = -1 \\ 10x + 6y = -10 \end{cases}$

7. $\begin{cases} -\dfrac{2}{3}x + y = \dfrac{1}{3} \\[2mm] \dfrac{1}{3}x - \dfrac{1}{2}y = \dfrac{1}{2} \end{cases}$

8. $\begin{cases} -6x + 4y = 4 \\ -3x - y = 26 \end{cases}$

9. $\begin{cases} \dfrac{1}{4}u + \dfrac{1}{2}v = \dfrac{3}{2} \\[2mm] \dfrac{1}{6}u + \dfrac{1}{3}v = 1 \end{cases}$

10. $\begin{cases} 3a + 7b = -8 \\ 2a + 3b = -2 \end{cases}$

11. $\begin{cases} -7c + 10d = 13 \\ 4c + 9d = 22 \end{cases}$

12. $\begin{cases} g + 2h = 3 \\ -5g + 4h = -8 \end{cases}$

13. $\begin{cases} 6a + 3b = -1 \\ 8a + 9b = 2 \end{cases}$

14. $\begin{cases} -0.05x + 0.05y = 0.15 \\ 0.1x - 0.1y = -0.3 \end{cases}$

15. $\begin{cases} 0.1x + 0.06y = 0.24 \\ 0.6x - 0.1y = -0.4 \end{cases}$

16. $\begin{cases} 9x + 4y = -3 \\ 3x + 12y = 7 \end{cases}$

Solve each of the following by any method.

17. $\begin{cases} x - y = -9 \\ y = -2x \end{cases}$

18. $\begin{cases} 9y = 7 - x \\ 9y + 2x = -13 \end{cases}$

19. $\begin{cases} 9x - 2y = -12 \\ 5x - 7y = 11 \end{cases}$

20. $\begin{cases} 8x + 7y = -11 \\ y + 2 = -2x \end{cases}$

MID-CHAPTER CHECK-UP

1. Solve the following systems of linear equations by graphing.

   a) $\begin{cases} y = \dfrac{2}{3}x \\ 2x - 3y = 6 \end{cases}$

   b) $\begin{cases} x + 3y = 9 \\ 2x - y = 4 \end{cases}$

2. Solve the following systems of linear equations by substitution.

   a) $\begin{cases} 3g + 4h = 2 \\ g = h + 3 \end{cases}$

   b) $\begin{cases} d = 3c - 4 \\ d = 1 - 2c \end{cases}$

3. Solve the following systems of linear equations by elimination.

   a) $\begin{cases} 2c - d = -4 \\ c + d = 1 \end{cases}$

   b) $\begin{cases} 3m + 4n = -1 \\ 2m - 5n = -3 \end{cases}$

3.5. Applications of Systems of Linear Equations

Objective: Solve application problems using system of linear equations.

Problem Solving Strategies and Tools (PSST)
When first looking at an application problem (or story problem), it is helpful to read the entire problem and then read it again more slowly to organize your thoughts. Note if additional information is needed.

A) Identify the unknown quantity and select a variable to represent it.

B) Write an equation or inequality that models the relationship between the known and unknown quantities.

C) Solve the equation or inequality. Check for reasonableness of solution.

D) Report the solution using a complete sentence.

Problem 1 : | *Worked Problem*

A chemist has 70 mL of a 50% methane solution.

a) How much of an 80% methane solution must be added so the final solution is 60% methane?

b) How much final solution will the chemist have?

Solution.

Using the Problem Solving Strategies and Tools:

1. *Identify the unknown quanties and select variables to represent them.*
Let x represent the amount of 80% methane solution
Let y represent the amount of 60% methane solution

2. *Write an equation that models the relationship between the known and unknown quantities.*
It is always a good idea to draw a picture to represent the situation.

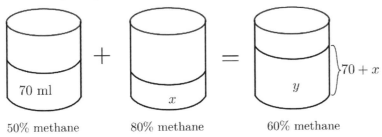

Set up the linear equations.
Equation 1: $70 + x = y$ (Amount of 50% methane solution + amount of 80% methane solutioin = amount of final solution)
Equation 2: $50\%(70 \text{ mL}) + 80\%(x) = 60\%(y)$ or $0.50(70) + 0.80x = 0.60y$

The system of linear equations: $\begin{cases} 70 + x = y \\ 0.50(70) + 0.80x = 0.60y \end{cases}$

3. *Solve the system of linear equations.*
   We will use substitution to solve the system of linear equations. Substitute first equation into the second.

$$0.50(70) + 0.80x = 0.60(70 + x) \quad \text{Distribute}$$
$$35 + 0.80x = 42 + 0.60x \quad \text{Subtract 35 from each side}$$
$$0.80x = 7 + 0.60x \quad \text{Subtract 0.60x from each side}$$
$$0.20x = 7 \quad \text{Divide each side by 0.20}$$
$$x = 35$$

Substitute $x=35$ into the first equation and solve for y.

$$70 + x = y \quad \text{Substitute } x=35$$
$$70 + 35 = y \quad \text{Add}$$
$$105 = y$$

4. *Report the solution.*
   The chemist needs 35 mL of the 80% methane solution. The chemist will have 105 mL of the final solution.

Problem 2 :	*Media/Class Example*

How many pounds of lima beans that cost 90¢ per pound must be mixed with   16 lb of corn that cost 50¢ per pound to make a mixture of vegetables that costs 65¢ per pound?

Problem 3 : | *You Try*

A coffee mix is to be made that sells for $2.50 by mixing two types of coffee. The first type of coffee has 40 mL of coffee that costs $3.00. How much of the second type of coffee that costs $1.50 should be mixed with the first?

Problem 4 : | *Media/Class Example*

How many grams of pure acid must be added to 40 grams of a 20% acid solution to make a solution which is 36% acid?

Problem 5 : | *You Try*

A solution of pure antifreeze is mixed with water to make a 65% antifreeze solution. How much of each should be used to make 70L of the 65% antifreeze solution?

Problem 6 : | *Worked Example*

Blondie's Candy shop has gourmet chocolate which sells for $12.00 a pound and fancy nuts which sell for $7.50 a pound. Blondie would like to sell a chocolate and nut mix for $10.50 a pound How much of each should she use to make 30 pounds of the new mixture?

Solution.

Using the Problem Solving Strategies and Tools:

1. *Identify the unknown quanties and select variables to represent them.*
   Let c represent the amount of gourmet chocolate
   Let n represent the amount of fancy nuts

2. *Write an equation that models the relationship between the known and unknown quantities.*
   It is always a good idea to draw a picture to represent the situation.

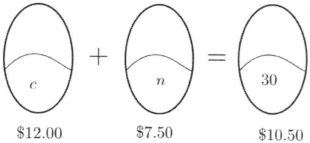

$$\$12.00 \qquad \$7.50 \qquad \$10.50$$

Set up the linear equations.
Equation 1: $c + n = 30$ (Amount of gourmet chocolate + amount of fancy nuts = amount of chocolate and nut mixture)
Equation 2: $\$12.00(c) + \$7.50(n) = \$10.50(30)$

The system of linear equations: $\begin{cases} c+n=30 \\ 12c+7.5n=315 \end{cases}$

3. *Solve the system of linear equations.*
    We will use elimination to solve the system of linear equations.

$$\begin{cases} c+n=30 \\ 12c+7.5n=315 \end{cases}$$ Multiply first equation by -12 to eliminate c

$$\begin{cases} -12c-12n=-360 \\ 12c+7.5n=315 \end{cases}$$ Add columns

$$-4.5n=-45$$ Divide each side by -4.5
$$n=10$$

Substitute $n = 10$ into the first equation and solve for c.

$$c+n=30 \quad \text{Substitute } n=10$$
$$c+10=30 \quad \text{Subtract 10 on each side}$$
$$c=20$$

4. *Report the solution.*
    Blondie needs 20 pounds of gourmet chocolate and 10 pounds of fancy nuts to make 30 pounds of chocolate and nut mixture that sells for $10.50 a pound.

Problem 7 : | *Media/Class Example*

The NorthStar Espresso Stand wants to offer a special blend of coffee beans. They sell Kona beans for $9.15 per pound and Colombian beans for $8.75 per pound. How many pounds of each type of beans should the barista combine to create 50 pounds of a blend that can be sold for $9.00 a pound?

Problem 8 : | *You Try*

A carpet manufacturer blends two fibers, one is 20% wool and the second is 50% wool. How many pounds of each fiber should be woven together to produce 600 lb of a fabric that is 28% wool?

Problem 9 : | *Media/Class Example*

After the baseball game, the teams go to *Sammy's Sweet Shop*. The *Frogs* order 6 waffle cones and 6 banan splits for $55.50. The *Monkeys* order 3 waffle cones and 9 banana splits for $57.75. How much does each item cost?

Problem 10 : | *You Try*

After the visit to the aquarium, the Mallards have lunch at a seafood bar. Fish and chips are $8.89 per order, a cup of chowder is $3.99 and soft drinks are $2.19 each. They purchased twice as many orders of fish and chips as cups of chowder and spent $87.08. How many orders of fish and chips and how many cups of chowder did they buy?

Problem 11 : | *Worked Example*

Mark and Jackie start from the same point and run in opposite directions. Mark runs 2 miles per hour faster than Jackie. After 3 hours, they are 30 miles apart. How fast did each run?

> **Solution.**

Using the Problem Solving Strategies and Tools:

1. *Identify the unknown quanties and select variables to represent them.*
   Let m represent Mark's speed in miles per hour
   Let j represent Jackie's speed in miles per hour

2. *Write an equation that models the relationship between the known and unknown quantities.*
   Note that distance = (rate)*(time) or $d = rt$
   Setting up a table is often useful for distance problems.

Person	Distance	Rate	Time
Mark	$3m$	m	3
Jackie	$3j$	j	3

   Draw a picture to represent the situation.

3. Set up the linear equations.
   Equation 1: $3m + 3j = 30$ (Mark's distance + Jackie's distance = 30 miles)
   Equation 2: $m = j + 2$ (Mark's runs 2 miles per hour faster than Jackie)

   The system of linear equations: $\begin{cases} 3m + 3j = 30 \\ m = j + 2 \end{cases}$

4. *Solve the system of linear equations.*
   We will use substitution to solve the system of linear equations. Substitute the second equation into the first.

$$3(j + 2) + 3j = 30 \quad \text{Distribute}$$
$$3j + 6 + 3j = 30 \quad \text{Combine like terms}$$
$$6j + 6 = 30 \quad \text{Subtract 6 from each side}$$
$$6j = 24 \quad \text{Divide each side by 6}$$
$$j = 4 \quad \text{Divide each side by } -4.5$$

   Substitute $j = 4$ into the second equation and solve for m.

$$m = j + 2 \quad \text{Substitute } j = 4$$
$$m = 4 + 2 \quad \text{Add}$$
$$m = 6$$

5. *Report the solution.*
   Mark ran at a rate of 6 miles per hour and Jackie at a rate of 4 miles per hour.

Problem 12 : | *Media/Class Example*

Joey leaves his home for school walking at a rate of 220 feet per minute. 15 minutes later, his mom realizes that Joey forgot his lunch. She gets on the bike, going at a rate of 1050 feet per minute. How long will it take Joey's mom to catch up with Joey? (Round answer to the nearest minute)

Problem 13 : | *You Try*

On a 130 mile family road trip to a mountain cabin, the Mallard's car traveled at an average speed of 60 mph on the highway and then reduced its speed to 40 mph when they drove over the snowy mountain pass. The trip took 2.5 hours. How long did the car travel at each speed?

Practice Problems: *Applications of Systems of Linear Equations*

Solve the following.

1. Mary and Bobby are 4-year old twins. Their combined weight is 75 lbs and their weight difference is 7 lbs (Bobby being the heavier one). How much do each of them weigh?

2. The day before a predicted snowstorm, a hardware store sold five times more flashlights than snow shovels. The store sold 138 of both items that day. How many of each were sold that day?

3. Mrs. Smith is bringing dessert for the holiday party. Her baking time is 15 minutes longer than her prep time. It will 65 minutes for the dessert to be ready. How long is the prep time and how long is the bake time?

4. How much pure alcohol must be added to 24 gallons of a 14% solution of alcohol in order to produce a 20% solution?

5. Mrs. Twinkle has tea that costs 26¢ an ounce that she will mix with 12 ounces of tea that cost 14¢ an ounce to form a new tea mix that will sell for 21¢ an ounce. How much 26¢ an ounce tea will Mrs. Twinkle use? How much will the new mixture weigh?

6. Blondie's Candy Shop has root beer barrels which sell for $15.00 a kilogram and butterscotch candies which sell for $6.50 a kilogram. Blondie would like to use 24 kilograms of butterscotch candies in the mix which she will sell for $9.00 a kilogram. How many kilograms of root beer barrels will she use? How much does the new mixture weigh?

7. A veterinarian told Barney that he needs to feed his dog with dog food that contains 11.3% fat. At the pet store, he can only find dog food with either 9% or 20% fat. Barney decides to mix his own dog food. If he wants 48 pounds of dog food, how many pounds of each pet store dog food does he need to buy?

8. Samantha's Syrup company manufactures some pure maple syrup and some syrup which is 85% maple syrup. How many quarts of each should she mix to make 150 quarts which is 96% maple syrup?

9. Two families visit the aquarium. The Smith family purchased 3 adult tickets and 2 youth tickets for $100.75. The Smiley family purchased an adult ticket and 4 youth tickets for $88.75. How much is one adult ticket and how much is one youth ticket?

10. Lily has exactly $5.06 to spend on trail mix for the algebra class hike and picnic. Chocolate candies cost 11¢ an ounce. Peanuts and raisins mix costs 22¢ an ounce. Lily wants 8 more ounces of peanuts and raisins mix than chocolate candies. How much of each will she be able to buy?

11. Wilma is making a platter for the algebra class picnic. A pound of ham costs twice as much as the pound of cheese. Wilma spends $48.75 on the platter and buys 5 pounds of each. How much is a pound of cheese and how much is a pound of ham?

12. There were 212 people at a play. Admission was $2.25 each for adults and 75¢ each for children. The total revenue from tickets was $283.50. How many children and how many adults attended?

13. The algebra students spend $90.44 to grill burgers and veggie dogs at their picnic. Each burger costs $2.27 and each veggie dog costs $1.92. The students want to grill twice as many burgers as veggie dogs. How many of each will they grill?

14. A passenger train and a freight train, on parallel tracks, start towards the same train station from two points 300 miles apart. If the rate of the passenger train exceeds the rate of the freight train by 15 miles per hour and they meet after 4 hours, what is the rate of each train?

15. Howard walks and jogs to campus each day. He averages 5km/hr (kph) walking and 9 kph jogging. The distance from home to campus is 8 km and Howard makes the trip in one hour. How long does he jog and how long does he walk?

16. As part of his flight training, Luke was required to fly to an airport and then return following the same route. The average speed to the airport was 90 mph and the average speed returning was 120 mph. The total flying time was 7 hours.

    a) Find how long it took to fly to the airport and how long was the return flight.

    b) Find the distance between the two airports.

17. Mark wants to jog and Jackie wants to rollerblade around Greenlake. Mark's average jogging speed is 4 mph slower than Jackie's average rollerblading speed. If both of them go in opposite direction starting at the community center and they first pass each other 20 minutes (or 0.2 hour) after they began, find each of their speeds.

18. 15 minutes (or $\frac{1}{4}$ hour) after Tommy got on the school bus, his mom realizes that Tommy forgot to bring his math homework. His mom gets in the car and drives at an average speed of 35 mph to try and catch up with the school bus. The school bus is going at an average speed of 20 mph. How long does it take for Tommy's mom to catch up to the school bus? (Round answer to the nearest whole minutes)

CHAPTER 3 ASSESSMENT

1. Solve each system of linear equations by the indicated method.

   a) By Graphing: $\begin{cases} x + y = 6 \\ y = 2x \end{cases}$

   b) By Substitution: $\begin{cases} x = 5y - 3 \\ 2x - 3y = 8 \end{cases}$

   c) By Elimination: $\begin{cases} 3x + 5y = -4 \\ 4x - 3y = -15 \end{cases}$

2. Solve each system of linear equations using any method.

   a) $\begin{cases} 9m - 3n = 12 \\ n = 3m - 4 \end{cases}$

   b) $\begin{cases} 5a + 2b = 3 \\ a - b = 2 \end{cases}$

   c) $\begin{cases} 4y = 3x + 4 \\ 6x - 8y = 4 \end{cases}$

   d) $\begin{cases} \dfrac{1}{2}x - \dfrac{1}{4}y = \dfrac{3}{2} \\ \dfrac{1}{3}x + \dfrac{1}{4}y = \dfrac{1}{6} \end{cases}$

3. Solve the following word problems.

   a) At one elementary school, there are twice as many girls as boys in the third grade. There are 108 third graders. How many boys and how many girls are there?

   b) A farmer wants to mix some 10% acid solution with some 30% acid solution to get 40 gallons of 15% acid solution. How many 10% and how mnay 30% acid solution should be mixed?

   c) Robin went to see her grandma who lives 63 miles away. She rode the bus travelling at 45 mph and then walked at 3 mph. The whole trip took 2.5 hours. How many miles did Robin walk?

CHAPTER 4
EXPONENTS

4.1. INTRODUCTION TO EXPONENTS

Objective: Introduce exponents and how to identify the base and its corresponding exponent.

Given the expression a^n, a is called the **base** and n is called the **exponent.**

Positive Exponents
If the exponent is a positive integer, then a^n tells us that the base, a, is multiplied n times, that is $a^n = \underbrace{a \cdot a \cdot a \ldots \ldots a}_{n \text{ times}}$.

Problem 1 :	*Worked Example*

Identify the base, exponent and simplify the expression 3^4.

   Solution.

The base is 3, the exponent is 4. That means we have to multiply the base, 3, four times or $3^4 = 3 \cdot 3 \cdot 3 \cdot 3 = 81$.

Problem 2 :	*Media/Class Example*

Identify the base, exponent and simplify the expression 2^5.

Problem 3 :	*You Try*

Identify the base, exponent and simplify the expression 4^2.

It is very important to be able to identify the base and its exponent accurately in order to simplify expressions correctly.

Problem 4 : | Worked Example

Simplify $4 \cdot 3^2$ and $(4 \cdot 3)^2$.

> **Solution.**

Following order of operation:

$$4 \cdot 3^2 = 4 \cdot 9 \quad \text{Exponent first} \qquad\qquad (4 \cdot 3)^2 = (12)^2 \quad \text{Parenthesis first}$$
$$= 36 \qquad\qquad\qquad\qquad\qquad\qquad = 144$$

Another way to look at this problem is to determine what is the base for the exponent 2. In the first expression, $4 \cdot 3^2$, only 3 is the base for the exponent 2. In the second expression, the product $(4 \cdot 3)$ is the base for the exponent 2.

$$4 \cdot 3^2 = 4 \cdot 3 \cdot 3 \qquad\qquad (4 \cdot 3)^2 = (4 \cdot 3)(4 \cdot 3)$$
$$= 36 \qquad\qquad\qquad\qquad\qquad = (12)(12)$$
$$= 144$$

Problem 5 : | Worked Example

Expand $4x^3$ and $(4x)^3$.

> **Solution.**

In this expression, $4x^3$, base 4 has an exponent 1 and base x has exponent 3. Therefore, $4x^3 = 4 \cdot x \cdot x \cdot x$.

In $(4x)^3$, the base is $(4x)$ and it has an exponent 3. $(4x)^3 = (4x)(4x)(4x)$ which can be rewritten as $64x^3$.

Problem 6 : | Media/Class Example

    a) Expand $-y^4$ and $(-y)^4$

                                        b) Expand $\dfrac{p^5}{2}$ and $\left(\dfrac{p}{2}\right)^5$

Problem 7 : | You Try

    a) Expand $(-3)^4$ and -3^4.   c) Expand $5 \cdot 2^3$ and $(5 \cdot 2)^3$

    b) Expand $\left(\dfrac{5}{6}\right)^2$ and $\dfrac{5^2}{6}$.   d) Expand $(-x)^2$ and $-x^2$

Zero and Negative Exponents

Let's take a look at what's happening to the expanded form and value of the expression as the base remains the same but the exponents decreases by 1.

1. Complete the table.

2. What do you notice?

Base 2	Expanded Form	Value
2^5	$=2 \cdot 2 \cdot 2 \cdot 2 \cdot 2$	$=32$
2^4	$=2 \cdot 2 \cdot 2 \cdot 2$	$=16$
2^3	$=2 \cdot 2 \cdot 2$	$=8$
2^2	$=2 \cdot 2$	$=4$
2^1		
2^0		

Base 2	Expanded Form	Value
2^0		
2^{-1}		
2^{-2}		
2^{-3}		
2^{-4}		
2^{-5}		

This suggests the following definitions:

Zero Property of Exponents: $a^0 = 1$, if $a \neq 0$.

Negative Property of Exponents:

If a is any real number such that $a \neq 0$ and n a positive integer then $a^{-n} = \dfrac{1}{a^n}$.

Let us see what happens if we change the base to 10.

Base 10	Expanded Form	Value
10^5	$=10 \cdot 10 \cdot 10 \cdot 10 \cdot 10$	$=100000$
10^4	$=10 \cdot 10 \cdot 10 \cdot 10$	$=10000$
10^3	$=10 \cdot 10 \cdot 10$	$=1000$
10^2	$=10 \cdot 10$	$=100$
10^1		
10^0		

Base 10	Expanded Form	Value
10^0		
10^{-1}		
10^{-2}		
10^{-3}		
10^{-4}		
10^{-5}		

Problem 10 : | *Media/Class Example*

Simplify the following.

a) 3^{-2}

c) $\dfrac{7}{2^0}$

b) -4^0

d) $5^0 + 5^{-1}$

Problem 11 : | *You Try*

Simplify the following.

a) 6^{-1}

c) $\dfrac{5^0}{2}$

b) $2 \cdot 5^0$

d) $2^0 - 2^{-3}$

Practice Problems: *Introduction to Exponents*

Insert >, < or = to make the statement true without using a calculator.

1. $5^{20} \boxed{} 5^{-20}$

5. $\left(\dfrac{1}{5}\right)^{20} \boxed{} 5^{-20}$

2. $-5^{20} \boxed{} (-5)^{20}$

3. $-5^{20} \boxed{} -5^{-20}$

6. $500^0 \boxed{} 5^0$

4. $-5^{20} \boxed{} \dfrac{1}{5^{20}}$

7. $5^{0.1} \boxed{} 5^0$

Expand the following expressions.

8. w^4

14. $\dfrac{5^3}{p}$

19. g^3

25. $\dfrac{g^5}{7}$

9. $(-w)^2$

20. $-g^3$

10. $-w^2$

15. $\dfrac{5^3}{p^3}$

21. $(-g)^3$

26. $\left(\dfrac{g}{-7}\right)^5$

11. $(2w)^4$

22. $7g^5$

27. $(k+2)^2$

16. $\dfrac{5}{p^3}$

12. $2w^4$

23. $(7g)^5$

28. $k^2 + 2^2$

17. $(x-7)^2$

13. $\left(\dfrac{5}{p}\right)^3$

18. $x^2 - 7^2$

24. $\dfrac{g}{-7^5}$

Simplify the following.

29. $2 \cdot 8^0$

36. 4^{-3}

30. $\left(\dfrac{3}{5}\right)^0$

37. $(-4)^3$

31. 5^2

38. -4^3

32. -5^2

39. $8^0 - 2^{-3}$

33. $(-5)^2$

40. $3^{-2} + \left(\dfrac{1}{3}\right)^0$

34. 5^{-2}

41. $19^0 - 6^{-1}$

35. 4^3

42. $\dfrac{3}{5}(7)^0 + \left(\dfrac{73}{5}\right)^0$

4.2. PROPERTIES OF EXPONENTS

Objective: Simplify expressions using the properties of exponents.

Expressions with exponents can often be simplified using a few basic exponent properties. Exponents represent repeated multiplication. We will use this fact to discover the important properties.

Problem 1 : | *Worked Example*

Simplify a^3a^2

> **Solution.**

$$
\begin{aligned}
& a^3a^2 && \text{Expand} \\
=& (a \cdot a \cdot a) \cdot (a \cdot a) && \text{Rewrite in exponent form} \\
=& a^5
\end{aligned}
$$

The outcome is the same if we added the exponents: $a^3a^2 = a^5$. We add the exponent when the bases are the same. This is known as the **product rule of exponents.**

$$\boxed{\textbf{Product Rule of Exponents: } a^m a^n = a^{m+n}}$$

Problem 2 : | *Worked Example*

Simplify $3^2 \cdot 3^6 \cdot 3$

> **Solution.**

$$
\begin{aligned}
& 3^2 \cdot 3^6 \cdot 3 && \text{Expand} \\
=& (3 \cdot 3) \cdot (3 \cdot 3 \cdot 3 \cdot 3 \cdot 3 \cdot 3) \cdot 3 && \text{Rewrite in exponent form} \\
=& 3^9 && \text{Our Solution}
\end{aligned}
$$

Problem 3 : | *Worked Example*

Simplify $(2x^3y^2) \cdot (5xy^2)$.

> **Solution.**

$$
\begin{aligned}
& (2x^3y^5) \cdot (5xy^2) && \text{Expand} \\
=& (2 \cdot x \cdot x \cdot x \cdot y \cdot y) \cdot (5 \cdot x \cdot y \cdot y) && \text{Multiply 2 and 5; Rewrite } x \text{ and } y \text{ in exponent form} \\
=& 10x^4y^7 && \text{Our Solution}
\end{aligned}
$$

Problem 4 : | *Media/Class Example*

Simplify the following.

    a) $p^5 p^7$
                                             b) $(3m^9 n^4) \cdot (3^2 m^2 n^6)$

Problem 5 : | *You Try*

Simplify the following.

a) $x^6 x^8 x^2$
                                         b) $(x^2 y) \cdot (x y^3)$

Next, we divide exponential expressions.

Problem 6 : | *Worked Example*

Simplify $\dfrac{a^5}{a^2}$

    Solution.

$$\frac{a^5}{a^2} \quad \text{Expand}$$

$$= \frac{aaaaa}{aa} \quad \text{Simplify; Note that } \frac{a}{a} = 1$$

$$= aaa \quad \text{Rewrite in exponent form}$$

$$= a^3 \quad \text{Our Solution}$$

The outcome is the same if we subtracted exponents, $\dfrac{a^5}{a^2} = a^{5-2} = a^3$. We subtract the exponent when the bases are the same. This is known as the quotient rule of exponents.

$$\boxed{\textbf{Quotient Rule of Exponents: } \frac{a^m}{a^n} = a^{m-n}, \ a \neq 0}$$

Problem 7 : | *Worked Example*

Simplify $\dfrac{c^3}{c^5}$

Solution.

$$\dfrac{c^3}{c^5} \qquad \text{Expand}$$

$$= \dfrac{c \cdot c \cdot c}{c \cdot c \cdot c \cdot c \cdot c} \qquad \text{Simplify; Note that } \dfrac{c}{c} = 1$$

$$= \dfrac{1}{c \cdot c} \qquad \text{Rewrite in exponent form}$$

$$= \dfrac{1}{c^2} \qquad \text{Our Solution}$$

Problem 8 : | *Worked Example*

Simplify $\dfrac{7^{13}}{7^2}$

Solution.

$$\dfrac{7^{13}}{7^5} = \dfrac{7 \cdot 7 \cdot 7 \cdot 7 \cdot 7 \cdot 7 \cdot 7 \cdot 7 \cdot 7 \cdot 7 \cdot 7 \cdot 7 \cdot 7}{7 \cdot 7 \cdot 7 \cdot 7 \cdot 7} \qquad \text{Expand and Simplify; Note that } \dfrac{7}{7} = 1$$

$$= 7 \cdot 7 \cdot 7 \cdot 7 \cdot 7 \cdot 7 \cdot 7 \cdot 7 \qquad \text{Rewrite in exponent form}$$

$$= 7^8 \qquad \text{Our Solution}$$

Problem 9 : | *Worked Example*

Simplify $\dfrac{5a^3b^2}{2ab^5}$

Solution.

$$\dfrac{5a^3b^2}{2ab^5} = \dfrac{5 \cdot a \cdot a \cdot a \cdot b \cdot b}{2 \cdot a \cdot b \cdot b \cdot b \cdot b \cdot b} \qquad \text{Expand and simplify}$$

$$= \dfrac{5 \cdot a}{2 \cdot b \cdot b} \qquad \text{Rewrite in exponent form}$$

$$= \dfrac{5a}{2b^2} \qquad \text{Our Solution}$$

Problem 10 : | *Media/Class Example*

Simplify the following.

a) $\dfrac{p^7}{p^4}$

b) $\dfrac{10m^{18}n^5}{5m^9n^6}$

Problem 11 : | *You Try*

Simplify the following.

a) $\dfrac{x^2}{x^{10}}$

b) $\dfrac{y^5z^3}{y^9z}$

A third property we will look at will have an exponent expression raised to another exponent.

Problem 12 : | *Worked Example*

Simplify $(a^2)^3$

> **Solution.**

$(a^2)^3$	Base, a^2 is raised to exponent 3; Multiply a^2 three times
$=a^2 \cdot a^2 \cdot a^2$	Expand a^2
$=(a \cdot a) \cdot (a \cdot a) \cdot (a \cdot a)$	Rewrite without parenthesis
$=a \cdot a \cdot a \cdot a \cdot a \cdot a$	Rewrite in exponent form
$=a^6$	Our solution

We end up multiplying the exponents, $(a^2)^3 = a^{2\cdot3} = a^6$. This is known as the power rule of exponents.

$$\boxed{\textbf{Power Rule of Exponents: } (a^m)^n = a^{mn}}$$

Problem 13 : | *You Try*

Simplify the following.

a) $\left(p^4\right)^5$
                                                    b) $\left(x^{10}\right)^2$

The next two properties are similar.

Problem 14 : | *Worked Example*

Simplify $(ab)^3$

   Solution.

   $(ab)^3$ Base, (ab) is raised to exponent 3; Multiply (ab) three times

   $= (ab) \cdot (ab) \cdot (ab)$ Three $a's$ and three $b's$ can be written with exponents

   $= a^3 b^3$ Our Solution

This is known as the power of a product rule.

$$\boxed{\textbf{Power of a Product Rule: } (ab)^m = a^m b^m}$$

It is important to be careful to only use the power of a product rule with multiplication inside parenthesis. A common error is to use a similar rule for a *sum*, or a *difference* raised to a power. There is no similar rule in those cases.

Warning. $(a \pm b)^m \neq a^m \pm b^m$ Beware of this error!!

Problem 15 : | *You Try*

Simplify the following.

a) $(2a)^5$
                                                    b) $\left(3x^2\right)^2$

What if there is division inside the parentheses instead of multiplication?

Problem 16 : | *Worked Example*

Simplify $\left(\frac{a}{b}\right)^3$

    Solution.

$$\left(\frac{a}{b}\right)^3 \qquad \text{Base,} \left(\frac{a}{b}\right) \text{is raised to exponent 3; Multiply} \left(\frac{a}{b}\right) \text{three times}$$

$$= \left(\frac{a}{b}\right)\left(\frac{a}{b}\right)\left(\frac{a}{b}\right) \qquad \text{Multiply fractions and rewrite using exponents}$$

$$= \frac{a^3}{b^3} \qquad \text{Our Solution}$$

This is known as the power of a quotient rule.

$$\boxed{\textbf{Power of a Quotient Rule: } \left(\frac{a}{b}\right)^m = \frac{a^m}{b^m}, b \neq 0}$$

Problem 17 : | *Media/Class Example*

Simplify the following.
a) $(3x)^2$

                                                  b) $\left(\frac{2}{m}\right)^4$

Problem 18 : | *You Try*

Simplify the following.
a) $(2x)^5$

                                                  b) $\left(\frac{w}{6}\right)^2$

The power of a power, product and quotient rules are often used together to simplify expressions.

Problem 19 : | Worked Example

Simplify $(x^3y^2)^4$

Solution.

$(x^3y^2)^4$ Raise the exponent of 4 on each factor, multiplying powers

$=x^{12}y^8$ Our solution

Problem 20 : | Worked Example

Simplify $\left(\dfrac{a^3}{c^8}\right)^2$

Solution.

$\left(\dfrac{a^3}{c^8}\right)^2$ Raise the exponent of 2 on each factor, multiplying powers

$=\dfrac{a^6}{c^{16}}$ Our Solution

Problem 21 : | Worked Example

Simplify $(4x^2y^5)^3$

Solution.

$(4x^2y^5)^3$ Raise the exponent of 3 on each factor, multiplying powers

$=4^3x^6y^{15}$ Evaluate 4^3

$=64x^6y^{15}$ Our Solution

Problem 22 : | Media/Class Example

Simplify the following.

a) $(3^2x^5)^2$

b) $\left(\dfrac{2p^5}{m^3}\right)^4$

Problem 23 : | *You Try*

Simplify the following.

a) $(2^2 x^7)^3$

b) $\left(\dfrac{w^3}{4m^2} \right)^2$

In this section we have discussed 5 different exponent properties. These rules are summarized in the following table.

Rules of Exponents

Product Rule	$a^m a^n = a^{m+n}$
Quotient Rule	$\dfrac{a^m}{a^n} = a^{m-n}, \ a \neq 0$
Power Rule	$(a^m)^n = a^{mn}$
Power of a Product Rule	$(ab)^m = a^m b^m$
Power of a Quotient Rule	$\left(\dfrac{a}{b} \right)^m = \dfrac{a^m}{b^m}, \ b \neq 0$

When simplifying an expression, it is often necessary to use more than one of these rules. There can be a bit of flexibility as to which property to use first. You still must be aware of order of operations.

Problem 24 : | *Media/Class Example*

Simplify the following.

a) $(x^3 y \cdot x y^2)^4$

b) $\dfrac{12(m^4)^2}{4m^6}$

Problem 25 : | *You Try*

Simplify the following.

a) $7a^3(2a^4)^3$

b) $\dfrac{(3m^3n)(10m^2n)}{2m^4n^2}$

Practice Problems: *Properties of Exponents*

Simplify.

1. $9 \cdot 9^4 \cdot 9^2$

2. $x^2 \cdot x^3 \cdot x$

3. $3x \cdot 4x^2$

4. $(3m)(4mn)$

5. $(2m^4n^2)(4nm^2)$

6. $x^2y^4 \cdot xy^2$

7. $(7^2)^4$

8. $(y^3)^4$

9. $(xy)^3$

10. $(2a^4)^4$

11. $(2xy)^4$

12. $(2u^3v^2)^2$

13. $\dfrac{p^5}{p^3}$

14. $\dfrac{3^2}{3}$

15. $\dfrac{8y^4}{4y}$

16. $\dfrac{3nm^2}{6n}$

17. $\dfrac{10x^2y^4}{5xy}$

18. $\dfrac{4x^3y^4}{3xy^3}$

19. $\dfrac{2xy^3}{4xy}$

20. $\dfrac{(3y^4)^2}{y^8}$

21. $\left(\dfrac{7m^{14}}{m^3}\right)^2$

22. $2x(x^4y^4)^4$

23. $\left(\dfrac{(2x)^3}{x^3}\right)^2$

24. $\dfrac{2a^2b^2}{(a^4b)^2}$

25. $\dfrac{x^2y \cdot (y^4)^2}{2y^4}$

26. $\dfrac{m^3(n^4)^2}{2mn}$

27. $(xy)^3$

28. $(2xy^3)^4$

4.3. NEGATIVE EXPONENTS AND ZERO POWER

Objective: Simplify expressions with negative exponents using the properties of exponents.

Zero Property of Exponents: $a^0 = 1$, if $a \neq 0$.

Problem 1 : | *Worked Example*

Simplify the following.

a) $(3x^2)^0, x \neq 0$

> **Solution.**
>
> $(3x^2)^0 = 1$; Base $(3x^2)$ is raised to a power 0 and by the definition of the zero property of exponent, the expression simplifies to 1.

b) $-5x^0$

> **Solution.**
>
> $-5x^0 = -5$; Exponent 0 applies only to base x. That means, $x^0 = 1$. However, -5 is multiplied by x^0. Therefore, $-5x^0 = -5(1) = -5$.

Problem 2 : | *Media/Class Example*

Simplify the following.

a) $4x^0$

c) $(-4)^0$

b) -4^0

d) $4^2 - 4^0$

Problem 3 : | *You Try*

Simplify the following.

a) $7(b^4)^0$

c) $\dfrac{2^0}{7}$

b) $\left(\dfrac{2}{7}\right)^0$

d) $\dfrac{7}{(b^4)^0}$

Negative Property of Exponents: If a is any real number such that $a \neq 0$ and n a positive integer, then

$$a^{-n} = \frac{1}{a^n}$$

(the negative exponent of a number is the reciprocal of the positive exponent of that number)

If you think of expression as fractions, factors in the numerator with negative exponents are moved to the denominator. Factors in the denominator with negative exponents are moved to the numerator. Once the factor is moved either to the numerator or denominator, the sign of the exponent changes. Remember that it is the sign of the exponent that changes and not the sign of the expression.

Warning. Negative exponents **never** make the bases negative.

World View Note: Nicolas Chuquet, the French mathematician of the 15th century wrote $12^{1\bar{m}}$ to indicate $12x^{-1}$. This was the first known use of the negative exponent.

Problem 4 : | *Media/Class Example*

Simplify the following. Write your answers with positive exponents only.

a) 4^{-2}

c) $5x^{-3}$

b) $\dfrac{1}{m^{-4}}$

d) $\dfrac{-7}{w^{-3}}$

Problem 5 : | *You Try*

Simplify the following. Write the answers with positive exponents only.

a) p^{-5}

c) $7w^{-1}$

b) $\dfrac{1}{5^{-2}}$

d) $-2c^{-2}$

In order to simplify an expression, we may use one property or a combination of the properties.

| **Problem 6 :** | *Worked Example* |

Simplify $\left(\dfrac{x^5}{x^{-3}}\right)^6$. Write the answer with positive exponents only.

Solution.

$$\left(\frac{x^5}{x^{-3}}\right)^6 \qquad \text{Apply Power of a Quotient Rule}$$

$$=\frac{(x^5)^6}{(x^{-3})^6} \qquad \text{Apply Power Rule}$$

$$=\frac{x^{30}}{x^{-18}} \qquad \text{Address Negative Exponent}$$

$$=x^{30}x^{18}$$

$$=x^{48} \qquad \text{Our Solution}$$

| **Problem 7 :** | *Media/Class Example* |

Simplify the following. Write the answers with positive exponents only.

a) -3^{-2}

c) $(3x^{-5})^{-3}$

b) $\dfrac{-3a^4}{b^{-7}}$

d) $\left(\dfrac{-2}{m}\right)^{-4}$

Problem 8 : | *You Try*

Simplify the following. Write the answers with positive exponents only.

a) $(-3)^4$

c) $(-2)^{-4}$

b) $\dfrac{-6}{n^{-5}}$

d) $\dfrac{b^{-4}}{a^{-8}}$

Problem 9 : | *Media/Class Example*

Simplify the following. Write the answers with positive exponents only.

a)

$$\dfrac{12ab^0}{4a^9b^{-5}}$$

c)

$$\dfrac{-15x^5}{30x^4}$$

d)

b)

$$-\left(\dfrac{2}{m}\right)^{-3}$$

$$\left(\dfrac{w^5}{-2w^{-4}}\right)^{-3}$$

Problem 10 : | *You Try*

Simplify the following. Write the answers with positive exponents only.

a) $\dfrac{-3x^5}{3x^5}$

c) $10^{14} \cdot 10^{-6}$

b) $p(2p)^{-5}$

d) $\left(\dfrac{w^5}{-2w^4}\right)^3$

Practice Problems: *Negative Exponents*

Simplify. Answers should contain only positive expontents.

1. -3^{-2}

2. $(-3)^{-2}$

3. $-x^{-4}$

4. $(-x)^{-4}$

5. $3w^{-5}$

6. $\dfrac{x^4}{x^{-16}}$

7. $\dfrac{r^{-4}}{r^{-10}r^5}$

8. $(y^5)^{-1}$

9. $\left(\dfrac{1}{4}\right)^{-1}$

10. $\left(\dfrac{5}{k^4}\right)^{-1}$

11. $\left(\dfrac{b^{-2}}{2}\right)^{-3}$

12. $\dfrac{4}{5}(10^{-3}p^7)^0$

13. $\dfrac{b^{15}b^{-7}}{b^3b^5}$

14. $\left(\dfrac{-2n^{-4}}{n^2}\right)^3$

15. $\left(\dfrac{3x^{-4}y^5}{12x^{10}y^3}\right)^{-2}$

16. $\dfrac{-20c^{-8}}{5c^3}$

17. $\dfrac{6(x^{-2})^3}{x^{-10}}$

18. $\left(\dfrac{2y^{-4}}{y^2}\right)^{-2}$

19. $\dfrac{(a^4)^2}{-2a^8}$

20. $\dfrac{(a^{-4})^2}{-2a^8}$

Insert the symbols >, < or = to make the statement true without using a calculator.

21. $(-3000)^{24}$ ☐ -3000^{24}

22. 3000^{24} ☐ 3000^{-24}

23. 3000^{0} ☐ 3000^{-1}

24. $(-3000)^{0}$ ☐ -30000^{0}

25. $\dfrac{1}{3000}$ ☐ 3000^{-1}

 Rescue Roody!

Roody was given the following problems to simplify. His answers were all incorrect. Help Roody figure out how to correctly simplify the expressions.

a) Simplify $\dfrac{3^{0}}{4}$

Roody's answer: 1. Roody says that any base, different from 0, and raised to the power 0 is 1.

b) Simplify 2^{-3}

Roody's answer: −6. Roody thinks that to simplify the expression, multiply the base, 2 and exponent, −3 to get the answer −6.

c) Simplify $-6x^{-3}$

Roody's answer: $\dfrac{1}{6x^{3}}$. Because 6 is negative and x has a negative exponent, Roody took the reciprocal of both factors and changed the negative sign to a positive sign.

4.4. SCIENTIFIC NOTATION

Objective: Understand scientific notation and convert between standard notation and scientific notation.

Scientific notation uses powers of 10 to represent very large or very small numbers. An example of a very large number is the distance that light travels in a year (approximately 6,000,000,000,000 miles). An example of a really small number is the mass of a hydrogen atom (approximately 0.00000000000000000000000166 grams).
Numbers in scientific notation have the following form:

$$\boxed{a \times 10^b, \text{ where } 1 \leqslant a < 10 \text{ and } b \text{ is an integer}}$$

The coefficient, a, must be greater than or equal to 1, and strictly less than 10. The exponent, b, tells us how many times we must multiply or divide by 10.

Problem 1 :	*Worked Example*

Convert 4.3×10^3 to standard notation.

Solution.

$$4.3 \times 10^3 = 4.3 \times \underbrace{10 \cdot 10 \cdot 10}$$

$$= 4.3 \times 1000$$

$$= 4300 \text{ (Standard Notation)}$$

In effect, we moved the decimal 3 places to the right, adding zeros as necessary. The positive exponent of base 10 indicates that the value of the entire number is larger than the coefficient, 4.3.

Problem 2 :	*Media/Class Example*

Convert 2×10^5 to standard notation.

How about when the exponent is negative?

Problem 3 : | *Worked Example*

Convert 5.7×10^{-4} to standard notation.

Solution.

$$5 \times 10^{-4} = 5 \times \frac{1}{10^4}$$

$$= 5 \times \frac{1}{10 \cdot 10 \cdot 10 \cdot 10}$$

$$= 5 \times \frac{1}{10,000}$$

$$= 5 \times 0.0001$$

$$= 0.00005$$

In effect, we moved the decimal 4 places to the left, introducing 0s as necessary. The negative exponent indicates that the value of the entire number is smaller than the coefficient 5.7.

Problem 4 : | *Media/Class Example*

Convert 6×10^{-1} to standard notation.

Problem 5 : | *You Try*

Convert the following from scientific notation to standard notation.

a) 3.42×10^6

c) 8×10^9

b) 7.85×10^{-3}

d) 1.9×10^{-7}

Now that we understand how to convert from scientific notation to standard notation, we will learn how to convert from standard notation to scientific notation.

Problem 6 : | *Worked Example*

Convert 123,000 to scientific notation.

Solution.

Our goal is to rewrite the number in the form $a \times 10^b$ where $1 < a < 10$ and b is an integer. Since 123,000 is larger than 1, we will do this by factoring out 10 until the "a" is between 1 and 10.

$$
\begin{aligned}
123,000 &= 12,300 \times 10 \\
&= 1,230 \times 10 \cdot 10 \\
&= 123 \times 10 \cdot 10 \cdot 10 \\
&= 12.3 \times 10 \cdot 10 \cdot 10 \cdot 10 \\
&= 1.23 \times 10 \cdot 10 \cdot 10 \cdot 10 \cdot 10 \\
&= 1.23 \times 10^5
\end{aligned}
$$

$123,000 = 1.23 \times 10^5$ in scientific notation.
In effect, we moved the decimal 5 places to the left. As the coefficient becomes smaller, the exponent must become larger, for the two numbers to remain equivalent.

Problem 7 : | *Worked Example*

Convert 0.0028 to scientific notation.

Solution.

Our goal is to rewrite the number in the form $a \times 10^b$ where $1 < a < 10$ and b is an integer. Since 0.0028 is smaller than 1, we will factor out $0.1 = \dfrac{1}{10} = 10^{-1}$ until the "a" is between 1 and 10.

$$
\begin{aligned}
0.0028 &= 0.028 \times 0.1 &&= 0.028 \times \frac{1}{10} \\[2mm]
&= 0.28 \times (0.1)(0.1) &&= 0.28 \times \frac{1}{10} \cdot \frac{1}{10} \\[2mm]
&= 2.8 \times (0.1)(0.1)(0.1) &&= 2.8 \times \frac{1}{10} \cdot \frac{1}{10} \cdot \frac{1}{10} \\[2mm]
&= 2.8 \times 0.001 &&= 2.8 \times \frac{1}{10^3} \\[2mm]
& &&= 2.8 \times 10^{-3}
\end{aligned}
$$

$0.0028 = 2.8 \times 10^{-3}$ in scientific notation.
In effect, we moved the decimal 3 places to the right. As the coefficient becomes larger, the exponent must become smaller, for the two numbers to remain equivalent.

Problem 8 : | *You Try*

Convert the following to scientific notation.

    a) 5,000,000,000   c) 0.001

    b) 0.0000762   d) 438

Scientific Notation and the Calculator

Calculators and computers display scientific notation in a slightly different way. Instead of displaying the "$\times 10$" portion of scientific notation, they display only the coefficient and the exponent of 10, labeled with an **E**.

For example, 2×10^{12} is displayed as **2 E 12** on a calculator, shown here.

3.1×10^{-4} is displayed as **3.1 E -4** on a calculator.

Problem 9 : | *Media/Class Example*

Using your calculator, perform the indicated operation. Write your answer in both scientific and standard notation.

    a) $5(3 \times 10^9)$

    b) $(2 \times 10^{-4})(3 \times 10^9)$

    c) $\dfrac{9 \times 10^7}{2 \times 10^3}$

Problem 10 : | *You Try*

Determine which value is larger and how many times larger than the other value.

a) 10^5 or 10^3

c) 10^{-2} or 10^{-1}

b) 10^{-2} or 10^{-4}

d) 10^{-1} or 10^2

Some wonderful sites that you may want to check out. Both sites take you on an adventure of magnitudes in powers of ten.

- https://www.youtube.com/watch?v=0fKBhvDjuy0
- http://htwins.net/scale2/

World View Note: Archimedes (287 BC - 212 BC), the Greek mathematician, developed a system for representing large numbers using a system very similar to scientific notation. He used his system to calculate the number of grains of sand it would take to fill the universe. His conclusion was 10^{63} grains of sand because he figured the universe to have a diameter of 10^{14} stadia or about 2 light years.

Practice Problems: *Scientific Notation*

Write each number in scientific notiation

1. 8,850

2. 0.081

3. 0.00000391

4. 0.000744

5. 1,090,000

6) 15,000,000,000

Write each number in standard notation

7. 8.7×10^5

8. 9×10^{-4}

9. 2×10^8

10. 2.56×10^2

11. 5.33×10^4

12. 6.7×10^{-5}

Complete the following table.

	Name	Standard Notation	Scientific Notation
13.	Trillion	1,000,000,000,000	
14.	Billion		1×10^9
15.	Million		1×10^6
16.	Thousand	1,000	
17.	Tenth	0.1	
18.	Hundredth	0.001	
19.	Millionth		1×10^{-6}
20.	Billionth		1×10^{-9}

Insert either < or > to make the statement true.

21. $2 \times 10^5 \boxed{} 2 \times 10^6$

22. $3 \times 10^5 \boxed{} 3 \times 10^{-5}$

23. $5 \times 10^{-5} \boxed{} 5 \times 10^{-6}$

24. $8 \times 10^5 \boxed{} 7 \times 10^5$

25. $2 \times 10^5 \boxed{} 2.1 \times 10^5$

26. $2 \times 10^6 \boxed{} 6 \times 10^2$

Perform the indicated operation. Write answer in scientific notation.

27. $(4.7 \times 10^5)(2 \times 10^{-3})$

28. $(3.1 \times 10^{-6})^2$

29. $1.5\,(2.3 \times 10^{-5})$

30. $\dfrac{4.8 \times 10^8}{2 \times 10^5}$

Word Problems.

31. As of Dec. 13, 2015, the national debt is nearing \$19 trillion. Write this amount in scientific notation. (source: www.usdebtclock.org)

32. It is estimated that internet traffic will grow to 88 exabytes per month in 2016 or 88,000,000,000,000,000,000 bytes. Write this amount in scientific notation. (source: www.cisco.com)

33. Some hummingbirds beat their wings at a rate of 80 times/second. This is 0.0125 second/beat. Write this amount in scientific notation. (source: www.hummingbirds.net)

34. In 2015, it is estimated that Mark Zuckerberg's shares in Facebook are worth $\$4.5 \times 10^{10}$. Zuckerberg wants to give 99% of his Facebook shares away. If Zuckerberg keeps 1% of his facebook shares, what this value? Write the answer in both scientific and standard notation.

35. A nanometer is 1×10^{-9} meters. A kilometer is 1×10^{3} meters. How many nanometers are in a kilometer?

36. For a typical 3 minute song, the mp3 file is 2.75×10^{6} bytes. How many mp3 files could be stored in 32 gigabytes? Write the answer in both scientific and standard notation. (1 gigabyte $= 1 \times 10^{9}$ bytes)

 **Rescue Roody!**

Help Roody figure out what he did wrong in each of these questions.

a) Rewrite 0.00325 in scientific notation. Roody's answer: 5.23×10^{-4}.

b) Rewrite 1.23×10^{-5} in standard notation. Roody's answer: 123,000.

CHAPTER 4 ASSESSMENT

Simplify the following expressions.

1. $(-9)^2$

2. -9^2

3. 9^{-2}

4. -9^{-2}

5. -9^0

6. $5x^0$

7. $\dfrac{3^0}{4}$

8. $\left(\dfrac{3}{4}\right)^0$

9. $3^{-2} - 4^0$

10. $\left(\dfrac{m^3}{7}\right)^2$

11. $(-2xy^2)(5xy)$

12. $\left(\dfrac{7}{w^2}\right)^{-1}$

13. $\dfrac{(n^{-3})^{-3}}{n}$

14. $8p^{-2}$

15. $\dfrac{6a^3}{4a^5}$

16. $\dfrac{-5n^3}{15n^2}$

Write the following in standard notation.

17. 2.345×10^8

18. 7.152×10^{-5}

Write the following in scientific notation.

19. $7,340,000,000,000$

20. 0.0000561

Perform the indicated operation. Write your answer in scientific notation.

21. $\dfrac{3.6 \times 10^4}{1.2 \times 10^{-2}}$

Chapter 5

Polynomials

5.1. Introduction to Polynomials

Objective: Introduce polynomials.

An **algebraic term** is a mathematical expression that contains constant numbers, as 2, 3, 5, etc, and variables, as x, y, multiplied together. The constants and variables can be part of a radical, and can have exponents.

Examples of algebraic terms are

$$3xy, \quad 5\sqrt{x}, \quad y^3\sqrt{5x}$$

An **algebraic expression** is made of several terms connected by addition or subtraction.

Examples of algebraic expressions are

$$3x + \sqrt{4x} - \frac{3}{x}, \quad 5y\sqrt{y} - 2y^5, \quad 5y - \frac{4}{5 - 3y^2}$$

A **polynomial** is a special kind of an algebraic expression. A polynomial is an algebraic expression that consists of

- a finite number of terms
- the exponents of all variables are positive integers

Note that this implies that no variables can be part of a denominator.

Examples of polynomials are

$$3x^2 - 7x + 5, \quad 6y^3 - 6, \quad \frac{3}{4}x^3 + 7x^2 + \frac{x}{3}$$

Problem 1 : | Worked Example

Determine whether the following is a polynomial or not. Explain why or why not.

1. $4x^2 - 5$
2. $y^4 + y^{-2} + 8$

Solution.

1. A polynomial because it is composed of a real number and product of a real number and a variable such that the variable has a positive integer exponent.

2. Not a polynomial because one of the variable exponent is not a positive integer.

Problem 2 : | Media/Class Example

Determine whether the following is a polynomial or not. Explain why or why not.

1. $2.5c^2 - c + 6$
3. $7g^{1/2}$

2. $\dfrac{5}{m+3}$
4. $\dfrac{p}{2}$

Problem 3 : | You Try

Determine whether the following is a polynomial or not. Explain why or why not.

1. $\sqrt{x+4}$
3. $g^2 + 4g - 2.3$

2. $5 - \dfrac{6}{h}$
4. $t^4 - \dfrac{5}{8}$

Algebraic terms can be connected by addition or subtraction. Polynomials are often defined by the number of terms in the expression. The following are the most common classification of polynomials.

Monomial: A singular term such as $7.4y$, $-3x^5$, xy^2, $\frac{1}{2}$

Binomial: Two unlike terms held together by addition or subtraction such as $1.3a - 4$, $6x^4 + \frac{5}{8}$, $m^2 - n^2$

Trinomial: Three unlike terms held together by addition, subtraction or both such as $x^2 + 6x + 8$, $9m^2 - m + \frac{4}{5}$, $a + b + c$

Expressions with more than three terms are simply called polynomials.

The **degree of a polynomial in one variable** is the highest exponent of that variable. For example, the degree of the polynomial, $6x^2 + 5x - 3$ is 2. The highest exponent for the variable x is 2.

Problem 4 : | *Media/Class Example*

Complete the table by writing the number of terms of the polynomial , degree of the polynomial and classifying the type of polynomial as monomial, binomial, etc.

Expression	Number of Terms	Degree	Type of Polynomial
$2 - 4c$			
$p^2 - \frac{1}{5}$			
$8m^4 + m^3 - 5m^2 - m + 1$			
$k + k^5 - \sqrt{3}$			
24			

Problem 5 : | *You Try*

Complete the table by writing the number of terms of the polynomial , degree of the polynomial and classifying the type of polynomial as monomial, binomial, etc.

Expression	Number of Terms	Degree	Type of Polynomial
$2m^4$			
-5.9			
$m^3 + 2m^2 - 6m + 7$			
$\frac{1}{8} - g^3$			
$7 - 1.5p + 0.2p^2$			

Practice Problems: *Introduction to Polynomials*

Determine whether the following expression is a polynomial or not.

1. $c^5 - 3c^4 + 8c - 1$

2. $b^3 - 9 + \dfrac{1}{b}$

3. $\dfrac{1}{2}w^2 + w - \dfrac{4}{5}$

4. $\sqrt{25x^2 + 9}$

5. π

6. $\dfrac{2+n}{n+7}$

Complete the table by writing the number of terms of the polynomial, degree of the polynomial and classifying the type of polynomial as monomial, binomial, etc.

	Expression	Number of Terms	Degree	Type of Polynomial
7.	$x^2 + 9.1x + 2.3$			
8.	$3n - 4n^4$			
9.	$-\dfrac{5}{8}$			
10.	$n^3 - 7n^2 + 15n - 20$			
11.	$2p$			
12.	$n^7 + n^3$			

5.2. ADDITION AND SUBTRACTION OF POLYNOMIALS

Objective: To add, subtract and evaluate polynomials.

Polynomials can be added or subtracted if there are like terms in the expression. What are like terms? They are terms whose variable and its exponent are the same. For example, $5x$ and $-2x$ have like terms because both terms have the same variable x with exponent 1. However, $5x$ and $2y$ are unlike terms because the variables in each term are not the same. $5x$ and $5x^2$ also have unlike terms because the exponent of the variable x are not the same even though both terms have the same variable, x.

Problem 1 :	*Media/Class Example*

Determine if the following terms are like or unlike. Explain your reasoning.

Terms	Like or Unlike	Explanation
$3y$ and y		
$4x^2$ and $-6x^2$		
$7w$ and $5w^3$		
$9xy$ and $4y$		
2 and z		

We will first add and subtract binomials. If the binomial has like terms, you add the coefficients and keep the same variable and its exponent.

Problem 2 :	*Worked Example*

Complete the table by finding the sum or difference of each binomial.

Given	Sum or Difference
$2x + 6x$	
$4x - 6x^2$	
$5y + 3z$	
$7x^3 - 8x^3$	

Solution.

Given	Sum or Difference	Explanation
$2x + 6x$	$=8x$	Like terms; add coefficients
$4x - 6x^2$	$=4x - 6x^2$	Unlike terms; exponents of variable are different
$5y + 3z$	$=5y + 3z$	Unlike terms; variables are different
$7x^3 - 8x^3$	$=-x^3$	Like terms; subtract coefficients

Problem 3 : | *Media/Class Example*

Complete the table by finding the sum or difference of each binomial and explaining what you did.

Given	Sum or Difference	Explanation
$9x^2 + x$		
$5x^2 - 4x^2$		
$2a + 8b$		
$\dfrac{3}{4}w - \dfrac{5}{6}w$		
$x^3 + \dfrac{2}{7}x^2$		

Problem 4 : | *You Try*

Complete the table by finding the sum or difference of each binomial.

Given	Sum or Difference	Explanation
$2x^2 - x^2$		
$x^2 - 3.5x$		
$5yz + 3z$		
$7x - 1.8$		
$\dfrac{2}{3}x + \dfrac{1}{6}x$		

We will now add and subtract polynomials. Remember that you can only add and subtract like terms. With a polynomial expression, we **cannot solve** for the variable but we can combine like terms. Generally, the answer is written in descending order with the highest power of the variable written first, then in order from greatest to least.

Problem 5 : | *Worked Example*

Perform the indicated operation: $(4x^2 - 2x + 8) + (3x^2 - 9x - 11)$ and write the answer in descending order.

Solution.

$$
\begin{aligned}
&(4x^2 - 2x + 8) + (3x^2 - 9x - 11) && \text{Remove parenthesis first} \\
&= 4x^2 - 2x + 8 + 3x^2 - 9x - 11 && \text{Combine like terms} \\
&= 7x^2 - 11x - 3 && \text{Our Solution}
\end{aligned}
$$

Problem 6 : | *Media/Class Example*

Perform the indicated operation: $(7x^2 + 2x^4 + 5x^3) + (6x^3 - 8x^4 - 9x^2)$ and write the answer in descending order.

Problem 7 : | *You Try*

Perform the indicated operation: $(6w + 8w^2) + (3 + 4w^2 - 3w)$ and write the answer in descending order.

When we subtract with polynomials, we must remember to distribute the subtraction sign to all terms within the parentheses. A subtraction sign before a parenthesis is the same as having -1 before the parenthesis. Subtraction will change all the signs within the parentheses.

Problem 8 : | *Worked Example*

Perform the indicated operation: $(5x^2 - 2x + 7) - (3x^2 + 6x - 4)$ and write the answer in descending order.

   Solution.

$$(5x^2 - 2x + 7) - (3x^2 + 6x - 4) \quad \text{Distribute negative through second parenthesis}$$
$$= 5x^2 - 2x + 7 - 3x^2 - 6x + 4 \quad \text{Combine like terms}$$
$$= 2x^2 - 8x + 11 \quad \text{Our Solution}$$

Problem 9 : | *Media/Class Example*

Perform the indicated operation: $(2x^3 - 3x^2 - 6x + 5) - (3x^3 - 9x^2 - 8x + 7)$ and write the answer in descending order.

Problem 10 : | *You Try*

Perform the indicated operation and write the answer in descending order.

a) $(21y^3 - 5y^2 + 7y + 5) - (-10y^3 - y^2 + 18y - 1)$

b) $(-7m^2 - 5m - 1) - (-m^2 - 8m + 6)$

Addition and subtraction of polynomials can be combined in the same problem.

Problem 11 : | *Worked Example*

Perform the indicated operation: $(2x^2 - 4x + 3) + (5x^2 - 6x + 1) - (x^2 - 9x + 8)$ and write the answer in descending order.

Solution.

$$
\begin{aligned}
&(2x^2 - 4x + 3) + (5x^2 - 6x + 1) - (x^2 - 9x + 8) && \text{Distribute negative through} \\
&= 2x^2 - 4x + 3 + 5x^2 - 6x + 1 - x^2 + 9x - 8 && \text{Combine like terms} \\
&= 6x^2 - x - 4 && \text{Our Solution}
\end{aligned}
$$

Problem 12 : | *Media/Class Example*

Perform the indicated operation: $(5m^2 + 2m - 7) - (m^2 - 9m + 1) + (3m^2 + m - 2)$ and write the answer in descending order.

Problem 13 : | *You Try*

Perform the indicated operation: $(y^2 - 3y - 4) + (5y^2 - 3y + 6) - (17y^2 - 8y - 1)$ and write the answer in descending order.

Practice Problems: *Addition and Substraction of Polynomials*

Combine like terms. Write answers in descending order.

1. $x^2 + 3x^2$

2. $2x^2 - x$

3. $6x^3 + 4x^2$

4. $5k^2 + 3k - 2k^2 + 6k$

5. $(x^2 + 5x^3) + (7x^2 + 3x^3)$

6. $(6x^3 + 5x) - (8x + 6x^3)$

7. $(4n^4 + 2n^2 + 3) + (2n^4 - 7n^2 - 4)$

8. $(4p^2 - 2p - 3) - (3p^2 - 6p + 3)$

9. $(4b^3 + 7b^2 - 3b) + (b^3 + 5b^2 + 8)$

10. $(7 + 4m + 8m^2) - (5m^2 + 6m + 1)$

11. $(4x^3 - 7x^2 + x) + (6x^3 + 7x^2 + 2x - 8)$

12. $(2 - 2n^2 + 7n^4) + (2 + 2n^2 + 4n^3 + 2n^4)$

13. $(7b^3 - b + 8) - (3b^3 + 7b^2 + 7b - 8) + (6b^2 - 3b + 3)$

14. $(3n^3 - 8n^2 - 1) + (7n^3 + 3n^2 - 6n + 2) + (4n^3 + 8n^2 + 7)$

15. $(8x^3 + 2x^2 - 2x + 9) - (3x^3 + 2x^2 - x + 1) - (5x^3 - x + 8)$

16. $(4x^3 + 7x^2 - 2x + 8) - (2x^3 - 6x^2 + 8) + (5x^3 - 4x^2 + 6x)$

 Rescue Roody!

Roody's is learning how to combine like terms. Unfortunately, none of his homework problems are correct. Help Roody.

a) Combine like terms: $x^2 - x$. Roody's work:
$$x^2 - x = x$$

b) Combine like terms: $(4x - 5) - (2x + 7)$. Roody's work:
$$(4x - 5) - (2x + 7) = 4x - 5 - 2x + 7$$
$$= 2x + 2$$

5.3. Multiplication of Polynomials

Objective: Multiply polynomials.
We will first look at the difference between adding or subtracting and multiplying two monomials.

Problem 1 : | *Worked Example*

Perform the indicated operation.

a) $(4x^3)(3x^2)$ 　　　　　　　　　　　　　b) $4x^3 + 3x^2$

Solution.

a) $(4x^3)(3x^2) = (4 \cdot x \cdot x \cdot x) \cdot (3 \cdot x \cdot x)$　　b) $4x^3 + 3x^2$ Unike terms;
　　　　　　　$= (4 \cdot 3) \cdot (x \cdot x \cdot x \cdot x \cdot x)$　　　　　　cannot combine terms
　　　　　　　$= 12x^5$

Problem 2 : | *Media/Class Example*

Perform the indicated operation.

a) $(5x^3)(-x^3)$ 　　　　　　　　　　　　　b) $5x^3 - x^3$

Problem 3 : | *You Try*

Perform the indicated operation.

a) $(x)(4x^2)$ 　　　　　　　　　　　　　c) $(6x^2)(-7x^2)$

b) $x + 4x^2$ 　　　　　　　　　　　　　d) $6x^2 - 7x^2$

Next, we consider multiplying a monomial by a polynomial. We have seen this operation before with distributing through parenthesis. Here we will see the exact same process.

Problem 4 : | *Worked Example*

Multiply $4x^3(5x^2 - 2x + 5)$

Solution.

$$4x^3(5x^2 - 2x + 5) \qquad \text{Distribute } 4x^3$$
$$= (4x^3)(5x^2) - (4x^3)(2x) + (4x^3)(5)$$
$$= 20x^5 - 8x^4 + 20x^3 \qquad \text{Our Solution}$$

Problem 5 : | *Class/Media Example*

Multiply the following.

a) $4x(3x+7)$

c) $-y(8y+9)$

b) $3n^2(6n+7)$

d) $5p^2(6p^2-2p+5)$

Problem 6 : | *You Try*

Multiply the following.

a) $-3(4r-7)$

c) $-3y^4(6y^3+2y-7)$

b) $(n+3)\cdot 2n^4$

d) $8w(w^2+4w-6)$

We will now consider multiplying two polynomials. This is done by multiplying each term in the first binomail by each term in the second binomial.

Problem 7 : | *Worked Example*

Multiply $(4x+7)(3x-2)$

Solution.

$$(4x+7)(3x-2) \quad \text{Distribute } (4x+7) \text{ to } (3x-2)$$
$$=4x(3x-2)+7(3x-2) \quad \text{Multiply}$$
$$=12x^2-8x+21x-14 \quad \text{Combine like terms}$$
$$=12x^2+13x-14 \quad \text{Our Solution}$$

In many textbooks multiplying two binomials is known as the FOIL method. The letters of FOIL help us remember every combination:

- **F** stands for First, we multiply the first term of each binomial
- **O** stand for Outside, we multiply the outside two terms.
- **I** stands for Inside, we multiply the inside two terms
- **L** stands for Last, we multiply the last term of each binomial.

Problem 8 : | *Worked Example*

Multiply $(4x+7)(3x-2)$ using the FOIL method.

Solution.

$$(4x+7)(3x-2) \quad \text{Use FOIL to multiply}$$
$$(4x)(3x)=12x^2 \quad F-\text{First terms } (4x)(3x)$$
$$(4x)(-2)=-8x \quad O-\text{Outside terms } (4x)(-2)$$
$$(7)(3x)=21x \quad I-\text{Inside terms } (7)(3x)$$
$$(7)(-2)=-14 \quad L-\text{Last terms } (7)(-2)$$

Therefore
$$(4x+7)(3x-2)$$
$$=12x^2-8x+21x-14 \quad \text{Combine like terms}$$
$$=12x^2+13x-14 \quad \text{Our Solution}$$

Problem 9 : | *Media/Class Example*

Multiply the following.

a) $(2x+1)(x-4)$

c) $(5y-6)(4y-1)$

b) $(2x+3y)(8x-7y)$

d) $(7n+6)(7n-6)$

Problem 10 : | *You Try*

Multiply the following.

a) $(x+8)(4x+8)$

c) $(5n+7)(n-9)$

b) $(4x+3y)(5x+8y)$

d) $(3n+2)(3n-2)$

A polynomial base raised to a positive integer exponent works exactly the same way as a variable base raised to a positive integer exponent. That is, the exponent tells us how many times to multiply the base, whether it is a variable or a polynomial.

Problem 11 : | *Worked Example*

Multiply $(2x+5)^2$

Solution.

Remember that exponent means repeated multiplication of the base. The base in this case is $(2x-1)$ and it is raised to the power 2. This means we have to multiply the base by itself or $(2x-1)(2x-1)$.

$$
\begin{aligned}
(2x+5)^2 &= (2x+5)(2x+5) \\
&= (2x)(2x)+(2x)(5)+(5)(2x)+5(5) \\
&= 4x^2+10x+10x+25 \\
&= 4x^2+20x+25
\end{aligned}
$$

Warning. $(2x+5)^2 \neq 4x^2+25$ as shown in the example above. The power rule for exponents does not apply when terms are added or subtracted.

Problem 12 : | *Worked Example*

Multiply $(y-2)^3$

Solution.

The base $(y-2)$ is raised to the power of 3. This means we have to multiply the base 3 times or $(y-2)(y-2)(y-2)$.

$$
\begin{aligned}
(y-2)^3 &= (y-2)(y-2)(y-2) \\
&= (y^2-2y-2y+4)(y-2) \\
&= (y^2-4y+4)(y-2) \\
&= y^2(y)+y^2(-2)-4y(y)-4y(-2)+4(y)+4(-2) \\
&= y^3-2y^2-4y^2+8y+4y-8 \\
&= y^3-6y^2+12y-8
\end{aligned}
$$

Warning. $(y-2)^3 \neq y^3-8$ as shown in the example above. The power rule for exponents does not apply when terms are added or subtracted.

Problem 13 : | *Worked Example*

Multiply $(a+b)^3$

Solution.

The base $(a + b)$ is raised to the power of 3. This means we have to multiply the base 3 times, that is, $(a+b)(a+b)(a+b)$. In the above example, we multiplied the first 2 factors first. Because multiplication is associative, multiplying the first two factors first yields the same product as multiplying the last 2 factors first. In this example, we will multiply the last 2 factors first.

$$
\begin{aligned}
(a+b)^3 &\quad \text{Expand} \\
=(a+b)(a+b)(a+b) &\quad \text{Multiply the last 2 factors first} \\
=(a+b)(a^2+ab+ba+b^2) &\quad \text{Multiplication is commutative so } ab=ba \\
=(a+b)(a^2+2ab+b^2) &\quad \text{Multiply term by term} \\
=a^3+2a^2b+ab^2+a^2b+2ab^2+b^3 &\quad \text{Combine like terms} \\
=a^3+3a^2b+3ab^2+b^3 &\quad \text{Our Solution}
\end{aligned}
$$

Warning. $(a+b)^3 \neq a^3+b^3$ as shown in the example above. The power rule for exponents does not apply when terms are added or subtracted.

Problem 14 : | *Media/Class Example*

Multiply the following.

a) $(3a-2b)^2$

c) $(y+1)^3$

b) $(2m-3)(4m^2+4m+5)$

d) $3(x+6)(2x-5)$

Problem 15 : | *You Try*

Multiply the following.

a) $(1-6n)^2$

c) $(6x+7)(2x^2+5x-1)$

b) $-2(x+3y)^2$

d) $(x-4)^3$

Problem 16 : | *Worked Example*

Perform the indicated operation: $(x + 5)(x - 5) - (x - 3)^2$ and write the answer in descending order.

Solution.

Rewrite $(x-3)^2=(x-3)(x-3)$ and remember to follow order of operation.

$$
\begin{aligned}
(x+5)(x-5)-(x-3)^2 &= (x+5)(x-5)-(x-3)(x-3) \\
&= (x^2-5x+5x-25)-(x^2-3x-3x+9) \\
&= (x^2-25)-(x^2-6x+9) \\
&= x^2-25-x^2+6x-9 \\
&= 6x-34
\end{aligned}
$$

Problem 17 : | *Media/Class Example*

Perform the indicated operation: $(2x + 1)^2 - (x - 4)(x + 4)$ and write the answer in descending order.

Problem 18 : | *You Try*

Perform the indicated operation: $2(x + 1)^2 - (2x - 3)^2$ and write the answer in descending order.

Practice Problems: *Multiplication of Polynomials*

Perform the indicated operation.

1. $6(p-7)$

2. $2x(6x+3)$

3. $5m^4(4m+4)$

4. $(c+3)(c-5)$

5. $(x+5)(x+3)$

6. $(3v-4)(5v-2)$

7. $(6x-7)(4x+1)$

8. $(x+3y)(3x+4y)$

9. $(a+b)(a-b)$

10. $(3-y)(3+y)$

11. $(5n-4)(5n+4)$

12. $(3p-7)^2$

13. $(a+b)^2$

14. $(w+2)^3$

15. $(c-1)^3$

16. $(a-b)^3$

17. $(r-7)(6r^2-r+5)$

18. $7(x-5)(x-2)$

19. $6(4x-1)(4x+1)$

20. $2(3n-2)(2n^2-2n+5)$

21. $3x(x-4)+(x-3)^2$

22. $(6-y)^2-2y(5y+4)$

23. $(x+5)^2-(x+6)(x-2)$

24. $(2x-1)^2-(x+1)^2$

 Rescue Roody!

Roody was told to perform the indicated operation but kept getting the wrong answer. Here are Roody's work. Help Roody.

a) $(x-9)^2=x^2-81$

b) $(x+8)^2=(x+8)(x+8)=x^2+64$

5.4. DIVISION OF POLYNOMIALS

Objective: Learn how to divide polynomials.

Dividing by a Monomial

To divide a polynomial by a monomial, we divide each term of the polynomial by the monomial and simplify the resulting expression.

We know that $\frac{12}{2}$ can be simplifed as 6. If $\frac{12}{2}$ was written instead as $\frac{2+4+6}{2}$, to simplify, we divide the denominator 2 into each term in the numerator, as such, $\frac{2}{2}+\frac{4}{2}+\frac{6}{2}=$ $1 + 2 + 3 = 6$. It is important to remember that each term in the numerator has to be divided by the denominator.

| **Problem 1 : \| *Worked Example*** |

Simplify $\dfrac{6x^4 - 18x^3 - 3x^2}{3x^2}$.

   Solution.

$$\dfrac{6x^4 - 18x^3 - 3x^2}{3x^2} \qquad \text{Divide each term in the numerator by } 3x^2$$

$$= \dfrac{6x^4}{3x^2} - \dfrac{18x^3}{3x^2} - \dfrac{3x^2}{3x^2} \qquad \text{Simplify each fraction}$$

$$= 2x^2 - 6x - 1 \qquad \text{Our Solution}$$

Verify that we have the correct soluton.

$$3x^2(2x^2 - 6x - 1) \qquad \text{Multiply quotient by divisor}$$
$$= 6x^4 - 18x^3 - 3x^2 \checkmark \qquad \text{Answer matches dividend}$$

Note. However many terms the polynomial you are dividing has, the resulting solution should also have the same number of terms. In the above example, the polynomial has four terms. Therefore, the solution should also contain four terms. A common mistake is to ignore the last term because the monomial goes evenly into the last term. Remember that $\dfrac{3x^2}{3x^2} = 1, \text{not } 0$.

Problem 2 : | *Media/Class Example*

Simplify $\dfrac{8x^3 + 4x^2 - 2x + 6}{4x^2}$.

Problem 3 : | *You Try*

Simplify the following:

a) $\dfrac{25k^3 + 15k^2 + 5k}{5k}$

b) $\dfrac{16m^3 + 12m^2 - 4m + 24}{12m^2}$

Long Division

Long division is required when we divide by polynomials which are not monomials. Long division with polynomials is very similar to long division with whole numbers.

Problem 4 : | *Worked Example*

Divide $631 \div 4$ using long division.

Solution.

$4\overline{)631}$ Rewrite problem as long division

$\begin{array}{r} 1 \\ 4\overline{)631} \\ \underline{-4} \\ 23 \end{array}$

   Divide 4 into 6

   Multiply quotient by divisor: $1 \cdot 4 = 4$ and subtract

   Bring down next number

$\begin{array}{r} 15 \\ 4\overline{)631} \\ \underline{-4} \\ 23 \\ \underline{-20} \\ 31 \end{array}$

   Repeat, divide 4 into 23

   Multiply quotient by divisor: $5 \cdot 4 = 20$ and subtract

   Bring down next number

$$157$$
$$4\overline{)631}$$
$$\underline{-4}$$
$$23$$
$$\underline{-20}$$

31	Repeat. Now divide 4 into 31
−28	Multiply quotient by divisor: 7·4=28 and subtract
3	This is the remainder

$$157+\frac{3}{4} \quad \text{Our Solution}$$

Verify that we have the correct solution.

157·4	Multiply integer part of quotient by divisor
=628	
628+3	Add the remainder to the above result
=631✓	Answer matches dividend

This same process will be used to divide polynomials. Be sure to write the polynomials in descending order before dividing.

Steps in Dividing Polynomials:

1. Divide first terms
2. Multiply the quotient by the divisor
3. Subtract
4. Bring down the next term
5. Repeat the whole process until no term can be brought down

Problem 5 : | *Worked Example*

Divide $\dfrac{x^2 + 2x - 12}{x + 5}$ using long division.

Solution.

$$\dfrac{x^2 + 2x - 12}{x + 5} \qquad \text{Rewrite problem as long division}$$

$$\begin{array}{r} x \\ x+5\overline{|x^2+2x-12} \\ \underline{-(x^2+5x)} \\ -3x-12 \end{array} \qquad \begin{array}{l} \text{Divide first terms: } \dfrac{x^2}{x} = x \\ \text{Multiply: } x(x+5) = x^2 + 5x \text{ then } \textbf{subtract} \\ \text{Bring down the next term} \end{array}$$

$$\begin{array}{r} x-3 \\ x+5\overline{|x^2+2x-12} \\ \underline{-(x^2+5x)} \\ -3x-12 \\ \underline{-(-3x-15)} \\ 3 \end{array} \qquad \begin{array}{l} \\ \\ \\ \text{Repeat, divide first terms: } \dfrac{-3x}{x} = -3 \\ \text{Multiply: } -3(x+5) = -3x - 15 \text{ then } \textbf{subtract} \\ \text{This is the remainder} \end{array}$$

$$x - 3 + \dfrac{3}{x+5} \qquad \text{Our Solution}$$

Verify that we have the correct solution.

$$\begin{array}{ll} (x-3) \cdot (x+5) & \text{Multiply polynomial part of quotient by divisor} \\ = x^2 + 5x - 3x - 15 & \text{Combine like terms} \\ = x^2 + 2x - 15 & \\ x^2 + 2x - 15 + 3 & \text{Add remainder part to above answer} \\ = x^2 + 2x - 12 \checkmark & \text{Answer matches dividend} \end{array}$$

Problem 6 : | *Media/Class Example*

Divide $\dfrac{x^2 - 3x - 53}{x - 9}$ using long division.

Problem 7 : | *You Try*

Divide $\dfrac{x^2 - 10x + 16}{x - 7}$ using long division.

Problem 8 : | *Worked Example*

Divide $\dfrac{3x^3 - 5x^2 - 32x + 7}{x - 4}$ using long division.

Solution.

$$\dfrac{3x^3 - 5x^2 - 32x + 7}{x - 4} \qquad \text{Rewrite problem as long division}$$

$$
\begin{array}{r}
3x^2 \\
x - 4\overline{)3x^3 - 5x^2 - 32x + 7} \\
\underline{-(3x^3 - 12x^2)} \\
7x^2 - 32x
\end{array}
$$

Divide first terms: $\dfrac{3x^3}{x} = 3x^2$

Multiply: $3x^2(x-4) = 3x^3 - 12x^2$ then **subtract**

Bring down the next term

$$
\begin{array}{r}
3x^2 + 7x \\
x - 4\overline{)3x^3 - 5x^2 - 32x + 7} \\
\underline{-3x^3 + 12x^2} \\
7x^2 - 32x \\
\underline{-(7x^2 - 28x)} \\
-4x + 7
\end{array}
$$

Repeat, divide first terms: $\dfrac{7x^2}{x} = 7x$

Multiply: $7x(x-4) = 7x^2 - 28x$ then **subtract**

Bring down the next term

$$
\begin{array}{r}
3x^2 + 7x - 4 \\
x - 4\overline{)3x^3 - 5x^2 - 32x + 7} \\
\underline{-3x^3 + 12x^2} \\
7x^2 - 32x \\
\underline{-7x^2 + 28x} \\
-4x + 7 \\
\underline{-(-4x + 16)} \\
-9
\end{array}
$$

Repeat, divide first terms: $\dfrac{-4x}{x} = -4$

Multiply: $-4(x-4) = -4x + 16$ then **subtract**

This is the remainder

$$3x^2 + 7x - 4 - \dfrac{9}{x - 4} \qquad \text{Our Solution}$$

Verify that we have the correct solution.

$$(3x^2 + 7x - 4) \cdot (x - 4) \qquad \text{Multiply polynomial part of quotient by divisor}$$
$$= 3x^3 - 12x^2 + 7x^2 - 28x - 4x + 16 \qquad \text{Combine like terms}$$
$$= 3x^3 - 5x^2 - 32x + 16$$
$$3x^3 - 5x^2 - 32x + 16 - 9 \qquad \text{Add remainder part to above answer}$$
$$= 3x^3 - 5x^2 - 32x + 7 \checkmark \qquad \text{Answer matches dividend}$$

Problem 9 : | *Media/Class Example*

Divide $\dfrac{6x^3 - 8x^2 + 10x + 103}{2x + 4}$ using long division.

Problem 10 : | *You Try*

Divide the following using long division.

a) $\dfrac{r^3 - r^2 - 16r + 8}{r - 4}$

b) $\dfrac{2x^3 + 12x^2 + 4x - 37}{2x + 6}$

If the polynomial has missing terms, we will put a zero for the coefficient of any missing terms. Make sure to write polynomials in descending order.

| **Problem 11 : | *Worked Example*** |

Divide $\dfrac{2x^3 + 42 - 4x}{x+3}$ using long division.

Solution.

$$\dfrac{2x^3 + 42 - 4x}{x+3} \qquad \text{Reorder dividend; } x^2 \text{ term missing, add } 0x^2$$

$$\begin{array}{r} 2x^2 \\ \hline x+3\,\big|\,2x^3 + 0x^2 - 4x + 42 \\ \underline{-(2x^3 - 6x^2)} \\ -6x^2 - 4x \end{array}$$

Divide first terms: $\dfrac{2x^3}{x} = 2x^2$

Multiply: $2x^2(x+3) = 2x^3 + 6x^2$ and subtract

Bring down the next term

$$\begin{array}{r} 2x^2 - 6x \\ \hline x+3\,\big|\,2x^3 + 0x^2 - 4x + 42 \\ \underline{-2x^3 - 6x^2} \\ -6x^2 - 4x \\ \underline{-(-6x^2 - 18x)} \\ 14x + 42 \end{array}$$

Repeat, divide first terms: $\dfrac{-6x^2}{x} = -6x$

Multiply: $-6x(x+3) = -6x^2 - 18x$ and subtract

Bring down the next term

$$\begin{array}{r} 2x^2 - 6x + 14 \\ \hline x+3\,\big|\,2x^3 + 0x^2 - 4x + 42 \\ \underline{-2x^3 - 6x^2} \\ -6x^2 - 4x \\ \underline{+6x^2 + 18x} \\ 14x + 42 \\ \underline{-(14x + 42)} \\ 0 \end{array}$$

Repeat, divide first terms: $\dfrac{14x}{x} = 14$

Multiply: $14(x+3) = 14x + 42$ and subtract

No remainder

$$2x^2 - 6x + 14 \qquad \text{Our Solution}$$

Verify that we have the correct solution.

$$(2x^2 - 6x + 14) \cdot (x+3) \qquad \text{Multiply polynomial part of quotient by divisor}$$
$$= 2x^3 + 6x^2 - 6x^2 - 18x + 14x + 42 \qquad \text{Combine like terms}$$
$$= 2x^3 - 4x + 42\checkmark \qquad \text{Answer matches dividend}$$

Problem 12 : | *Class/Media Exampled*

Divide $\dfrac{x^3 + 22 - 46x}{x + 7}$ using long division.

Problem 13 : | *You Try*

Divide $\dfrac{4n^3 + 22n^2 - 38}{4n + 6}$ using long division.

Practice Problems: *Division of Polynomials*

Perform the indicated operation.

1. $\dfrac{9m^4 + 18m^3 + 27m^2}{9m^2}$

2. $\dfrac{9x^4 + 24x^3 - 6x^2 - 3x}{3x}$

3. $\dfrac{20x^4 + x^3 + 2x^2}{4x^3}$

4. $\dfrac{20n^4 + 50n^3 + 10n^2 + 40n}{10n^2}$

5. $\dfrac{3k^3 + 4k^2 + 2k + 8}{8k}$

6. $\dfrac{5x^4 + 45x^3 + 4x^2 + 9x}{9x}$

7. $\dfrac{a^2 - 4a - 30}{a - 8}$

8. $\dfrac{v^2 - 2v - 80}{v - 10}$

9. $\dfrac{n^2 + 7n + 5}{n + 4}$

10. $\dfrac{6x^2 + 5x - 6}{3x - 2}$

11. $\dfrac{4r^2 + 8r + 5}{2r + 5}$

12. $\dfrac{x^3 - 16x^2 + 71x - 56}{x - 8}$

13. $\dfrac{k^3 - 4k^2 - 6k + 4}{k - 1}$

14. $\dfrac{n^2 - 4}{n - 2}$

15. $\dfrac{9v^2 + 2}{3v - 1}$

16. $\dfrac{a^3 - 15a - 22}{a + 2}$

17. $\dfrac{4m^3 - 13m^2 - 9}{m - 3}$

18. $\dfrac{2n^3 + 11n^2 - 18}{2n + 3}$

CHAPTER 5 ASSESSMENT

Perform the indicated operation and write answers in descending order. Then identify the degree of the resulting polynomial and state whether it is a monomial, binomial, trinomial, etc.

1. $(4x^2 + 3x - 7) + (x^2 - 3x + 9)$

2. $(7n^3 + 2n^2 - 9n + 8) - (3n^3 - 2n^2 - 9n + 4)$

3. $(2y^2 - 7y + 8) - (6y^2 + 6y - 8) + (4t^2 - 2y + 3)$

4. $2c^2(3c - 8)$

5. $(m + 8)(m - 7)$

6. $(2w - 5)(3w - 4)$

7. $\left(h + \dfrac{3}{4}\right)\left(h - \dfrac{3}{4}\right)$

8. $(p - 5)^2$

9. $(y + 4)(2y^2 + 2y - 5)$

10. $3(2x + 1)(x^2 + 6x - 1)$

Divide the following expressions.

11. $\dfrac{6m^2 + 12m - 9}{3m}$

12. $\dfrac{3x^2 + x - 10}{x + 2}$

13. $\dfrac{15x^3 - 10x^2 + 5x}{10x^2}$

14. $\dfrac{2x^3 - 9x^2 - 17x + 39}{2x - 3}$

CHAPTER 6
FACTORING

6.1. FACTORING GREATEST COMMON FACTOR AND BY GROUPING

Objective: Factor polynomials by finding the greatest common factor and by grouping.

Factoring is the reverse process of multiplication. Let us review multiplying a polynomial by a monomial.

Problem 1 :	Worked Example

Multiply: $3x(5x+8)$

Solution.

$$3x(5x+8) \quad \text{Multiply } 3x \text{ using the distributive property}$$
$$= 3x(5x)+3x(8) \quad \text{Rewrite each term}$$
$$= 15x^2+24x \quad \text{Our solution}$$

Factoring the Greatest Common Factor

To factor out the greatest common factor, ask yourself: What factors are in common?

Problem 2 :	Worked Example

Identify the greatest common factor (GCF) and completely factor the expression: $5a^3 + 15a$.

Solution.

First rewrite each term by finding the prime factors of each coefficient and expanding the variable.

$$\text{Factors of } 5a^3: \quad 5 \cdot a \cdot a \cdot a \qquad \text{Factors of } 15a: \quad 3 \cdot 5 \cdot a$$

Both terms, $5a^3$ and $15a$, have a greatest common factor $5 \cdot a = 5a$. Let us now factor the expression.

$$5a^3 + 15a = 5 \cdot a \cdot a \cdot a + 3 \cdot 5 \cdot a$$
$$= 5 \cdot a(a \cdot a + 3)$$
$$= 5a(a^2+3)$$

Check to make sure expression is factored correctly by applying the distributive property.

$$5a(a^2+3) = 5a^3+15a \qquad \text{Check!}$$

Problem 3 : | *Media/Class Example*

Identify the greatest common factor (GCF) and factor each expression completely.

a) $3y^2 + 12y - 24$

c) $12x^2 - 4x^3 + 6x^5$

b) $8h - 24h^2$

d) $5ab^2 + 10a^2b^2 + 15a^2b$

Problem 4 : | *You Try*

Identify the greatest common factor (GCF) and factor each expression completely. Be sure to check your answer.

a) $15m + 40$

c) $21a^3 + 14a^2 + 7a$

b) $14y^5 - 49y^2$

d) $15mn + 12m - 3mn^2$

World View Note: The first recorded algorithm for finding the greatest common factor comes from Greek mathematician Euclid around the year 300 BC!

When a polynomial has a negative leading coefficient, we usually factor out a negative sign, making the GCF a negative number.

Problem 5 : | *Worked Example*

Factor $-4a + 12b - 8c$ completely.

   Solution.

   Since the leading coefficient of the given polynomial is negative and 4 is a common factor, the given expression's GCF is -4. Factoring -4 out, we get:
$$-4a + 12b - 8c = -4(a - 3b + 2c)$$
   Notice the signs of the polynomial inside the parenthesis are opposite that of the original polynomial.

Check by using the distributive property.

$$-4(a - 3b + 2c) = -4a + 12b - 8c \qquad \text{Check!}$$

Problem 6 : | *Worked Example*

Factor $-x^2+5x-2$ completely.

Solution.

Since the leading coefficient of the given polynomial is negative and there are no other common factors, its GCF is -1. The polynomial factors as:

$$-x^2+5x-2 = -1(x^2-5x+2)$$
$$= -(x^2-5x+2)$$

Check to verify factor is correct.

$$-(x^2-5x+2) = -x^2+5x-2 \qquad \text{Check!}$$

Problem 7 : | *Media/Class Example*

Identify the greatest common factor (GCF) and completely factor the expression.

a) $-x-2y+3$ 　　　　　　　　　　　b) $-18p^3+12p^2-6p$

Problem 8 : | *You Try*

Identify the greatest common factor (GCF) and completely factor the expression.

a) $-15x^3+9x^2$ 　　　　　　　　　　b) $-6xyz+12xz-15xy$

Greatest common factor is not always a monomial. Sometimes it can be a binomial or any other polynomial.

Problem 9 : | *Worked Example*

Factor $4x(x-3)+7(x-3)$ completely.

Solution.

Let us take a look at the factors of each term.

Factors of first term, $4x(x-3)$: $4 \cdot x \cdot (x-3)$

Factors of second term, $7(x-3)$: $7 \cdot (x-3)$

Both terms have a greatest common factor $(x-3)$. Factoring $(x-3)$ out, we get:

$$4x(x-3)+7(x-3) = (x-3)(4x+7)$$

Check to verify factor is correct. From the factored expression we get:

$$(x-3)(4x+7) = x(4x+7) - 3(4x+7)$$
$$= 4x^2 + 7x - 12x - 21$$
$$= 4x^2 - 5x - 21$$

From the given expression, we get:

$$4x(x-3) + 7(x-3) = 4x^2 - 12x + 7x - 21$$
$$= 4x^2 - 5x - 21$$

The factored form and the given expression yield the same result: $4x^2 - 5x - 21$. Therefore, $(x-3)(4x-7)$ must be the correct factor for the given expression.

Problem 10 : | *Media/Class Example*

Identify the greatest common factor (GCF) and completey factor the expression.

a) $a^2(a+5) + 9(a+5)$ 　　　　　　　　　b) $10x^3(2x-3y) - 15x^2(2x-3y)$

Problem 11 : | *You Try*

Identify the greatest common factor (GCF) and completely factor the expression.

a) $8(t-3) - 3t(t-3)$ 　　　　　　　　　b) $a(a+7) + 5(a+7)$

Factoring by Grouping

Sometimes, there are no common factors to all the terms of the polynomial but there may be factors common to some of the terms. Let us see how the grouping technique helps to factor the expression.

Problem 12 : | *Worked Example*

Factor $10ab + 15b + 4a + 6$ completely.

Solution.

The expression has no common factor to all of its terms. Let us use grouping technique to factor by grouping the first two terms and the last two terms.

$$\underbrace{10ab + 15b} + \underbrace{4a + 6} \qquad \text{Group first 2 terms and last 2 terms}$$

$$= 5b(2a+3) + 2(2a+3) \qquad \text{Find GCF for each group and factor}$$

$$= (2a+3)(5b+2) \qquad \text{Factor out GCF, } (2a+3)$$

Check that factors are correct.

$$
\begin{aligned}
(2a+3)(5b+2) &= 2a(5b+2)+3(5b+2) \\
&= 10ab+4a+15b+6 \\
&= 10ab+15b+4a+6 \qquad \text{Check!}
\end{aligned}
$$

Problem 13 : | *Worked Example*

Factor $5xy-8x-10y+16$ completely.

> **Solution.**

The expression has no common factor to all its terms. Let us factor by grouping the first two terms and the last two terms. Note that the leading coefficient of the last two terms is negative. Its GCF must be a negative number.

$$\underbrace{5xy-8x}\ \underbrace{-10y+16} \qquad \text{Group first 2 terms and last 2 terms}$$

$$= x(5y-8)-2(5y-8) \qquad \text{Find GCF for each group and factor}$$

$$= (5y-8)(x-2) \qquad \text{Factor out GCF, } (5y-8)$$

Check that factors are correct.

$$
\begin{aligned}
(5y-8)(x-2) &= 5y(x-2)-8(x-2) \\
&= 5xy-10y-8x+16 \\
&= 5xy-8x-10y+16 \qquad \text{Check!}
\end{aligned}
$$

Problem 14 : | *Worked Example*

Factor $6x^3-15x^2+2x-5$ completely.

> **Solution.**

The expression has no common factors to all its terms. Let us factor by grouping the first two terms and the last two terms. Notice that the last two terms have no common factor. In this case, we will factor out a 1.

$$\underbrace{6x^3-15x^2}+\underbrace{2x-5} \qquad \text{Group first 2 terms and last 2 terms}$$

$$= 3x^2(2x-5)+1(2x-5) \qquad \text{Find GCF for each group and factor}$$

$$= (2x-5)(3x^2+1) \qquad \text{Factor out GCF, } (2x-5)$$

Check that factors are correct.

$$
\begin{aligned}
(2x-5)(3x^2+1) &= 2x(3x^2+1)-5(3x^2+1) \\
&= 6x^3+2x-15x^2-5 \\
&= 6x^3-15x^2+2x-5 \qquad \text{Check!}
\end{aligned}
$$

Problem 15 : | *Media/Class Example*

Factor each expression completely.

a) $xy + 2x + 4y + 8$

c) $49y^2 - 14y - 14y + 4$

b) $ab - 3b + 7a - 21$

d) $4ab + 12a - b - 3$

Problem 16 : | *You Try*

Factor each expression completely.

a) $xy + 2y + 6x + 12$

c) $w^3 - 5w^2 - w + 5$

b) $3ax + 21x - a - 7$

d) $12y^2 - 21y + 20y - 35$

Practice Problems: *Factoring the Greatest Common Factor and by Grouping*

Factor each expression completely.

1. $45x^2 - 25$

2. $56 - 35p$

3. $50x - 80y$

4. $7ab - 35a^2b$

5. $-3a^2b + 6a^3b^2$

6. $-32n^3 + 32n^2 + 8n$

7. $-5x^2 + 25x^3 - 15x^4$

8. $21p^2 + 30p + 27$

9. $-10x^4 + 20x^2 + 12x$

10. $30b^2 + 5ab - 15a^2$

11. $-27x^2y^2 + 12xy^2 - 9y^2$

12. $3x^3 + 15x^2 + 2x + 10$

13. $3n^3 - 2n^2 - 9n + 6$

14. $40r^3 - 8r^2 - 25r + 5$

15. $15b^3 + 21b^2 - 35b - 49$

16. $7xy - 49x + 5y - 35$

17. $16xy - 56x + 2y - 7$

18. $7n^3 + 21n^2 - 5n - 15$

19. $3mn - 8m + 15n - 40$

20. $8xy + 56x - y - 7$

21. $28p^3 + 21p^2 + 20p + 15$

22. $14v^3 + 10v^2 - 7v - 5$

23. $4y^3 - 12y^2 + 8y - 24$

24. $30x^3 - 20x^2 - 6x + 4$

25. $14u^2 + 42u + 4uv + 12v$

6.2. FACTORING $ax^2 + bx + c$ - PART I

Objective: Factor binomials and trinomials where the coefficient of the squared term is one.

Over the next two sections, we will learn how to factor trinomials of the form

$$ax^2 + bx + c.$$

This type of trinomial is known as a **quadratic** because it is of degree 2.

- ax^2 is known as the **quadratic term**

- bx is referred to as the **linear term**

- c is the **constant term**

In this section, we will focus on trinomials where $a = 1$. The next section will cover quadratics where $a \neq 1$ after factoring the greatest common factor.

Since factoring is the reverse process of multiplication, let us review that polynomial operation. Let us multiply $(x+6)(x+4)$.

$$
\begin{aligned}
&(x+6)(x+4) &&\text{Multiply the two binomials} \\
&= x(x+4) + 6(x+4) &&\text{Distribute each monomial} \\
&= x^2 + 4x + 6x + 24 &&\text{Combine like terms} \\
&= x^2 + 10x + 24 &&\text{Our Solution}
\end{aligned}
$$

There are some things to notice about our answer:

1. The answer is a trinomial. Therefore, if a quadratic trinomial can factor, the factors will be two binomials.

2. x^2 is the product of the first terms of each binomial.

3. b is the sum of the constants found in each binomial. From the above example, $10 = 4 + 6$.

4. c is the product of the constants. From the above example, $24 = 6(4)$.

Consequently, we need to find:

- factors of c
- that sum to b

Problem 1 : | Worked Example

Factor $x^2+8x+15$ completely.

Solution.

Let us get started with things we know. If this is factorable, the answer will be the product of two binomials. Since x^2 can only factor as $x \cdot x$, we can write
$$x^2+8x+15=(x \qquad)(x \qquad)$$
Next, we need find factors of 15 that sum to 8. Since 8 and 15 are both positive, we only consider positive factors.

Positive Factors of 15	Product of Factors	Sum of Factors
1, 15	$1 \cdot 15 = 15$	$1+15=16$
3, 5	$3 \cdot 5 = 15$	$3+5=8$

The last pair gives us what we need. We put those values into the binomials to get our final answer.
$$x^2+8x+15=(x+3)(x+5)$$
To check our answer, we perform multiplication and confirm we get back the original binomial.

$$\begin{aligned}(x+3)(x+5) &= x(x+5)+3(x+5)\\ &= x^2+5x+3x+15\\ &= x^2+8x+15 \qquad \text{Confirmed!}\end{aligned}$$

Problem 2 : | Worked Example

Factor $x^2+8x-20$ completely.

Solution.

We need to find factors of -20 that sum up to 8. Given that the product is negative, we have to consider combinations of positive and negative factors.

Factors of -20	Product of Factors	Sum of Factors
1, -20	$1 \cdot (-20) = -20$	$1+(-20)=-19$
2, -10	$2 \cdot (-10) = -20$	$2+(-10)=-8$

Notice that the last pair comes very close but the sign of the sum is the opposite of what we want. This means we need to swap the signs of the factors. Instead, use -2 and 10.
$$x^2+8x-20=(x-2)(x+10)$$
Check:

$$\begin{aligned}(x-2)(x+10) &= x(x+10)-2(x+10)\\ &= x^2+10x-2x-20\\ &= x^2+8x-20 \qquad \text{Confirmed!}\end{aligned}$$

Problem 3 : | *Worked Example*

Factor $x^2 + 5x + 9$ completely.

  Solution.

We need to find factors of 9 that sum up to 5. Since both 9 and 5 are positive, we only consider positive factors.

Positive Factors of 9	Product of Factors	Sum of Factors
1, 9	$1 \cdot 9 = 9$	$1 + 9 = 10$
3, 3	$3 \cdot 3 = 9$	$3 + 3 = 6$

Those are all the positive factors of 9 and none of them sum up to 5. A polynomial that is not factorable is called **prime**. Therefore, we say that $x^2 + 5x + 9$ is prime.

Problem 4 : | *Media/Class Example*

Factor the following trinomials completely, if possible. If not factorable, identify it as prime.

  a) $x^2 - 4x + 3$ b) $y^2 + 9y - 10$ c) $m^2 + 4m + 14$

Problem 5 : | *You Try*

Factor the following trinomials completely, if possible. If not factorable, identify it as prime.

  a) $a^2 + 9a + 18$ b) $y^2 - 7y + 6$

  c) $y^2 - 7y - 18$ d) $x^2 - 4x - 12$

  e) $c^2 - 10c + 20$ f) $p^2 + 13p + 36$

  g) $m^2 + 2m - 3$ h) $n^2 - 5n - 6$

Let us summarize what we see happening with the signs .

1. When the constant term of the trinomial is positive, its factors must have the same sign.

   - If the linear term is positive, trinomial factors as $(\_ + \_)(\_ + \_)$. For example, $x^2 + 8x + 15 = (x+3)(x+5)$.

   - If the linear term is negative, trinomial factors as $(\_ - \_)(\_ - \_)$. For example, $x^2 - 8x + 15 = (x-3)(x-5)$.

2. When the constant term of the trinomial is negative, then its factors must have different signs. The trinomial factors as $(\_ + \_)(\_ - \_)$. The sign of the linear term will indicate which factor is positive and which is negative. For example:

$$x^2 + 2x - 15 = (x-3)(x+5)$$
$$x^2 - 2x - 15 = (x+3)(x-5)$$

Problem 6 :	*Worked Example*

Factor $x^2 - 10x + 25$ completely.

Solution.

We need to find factors of 25 that sum up to -10. Since the constant term, 25, is positive and the linear term has a negative coefficient, we will consider negative factors only.

Negative Factors of 25	Product of Factors	Sum of Factors
$-1 - 25$	$(-1) \cdot (-25) = 25$	$(-1) + (-25) = -26$
$-5, -5$	$(-5) \cdot (-5) = 25$	$(-5) + (-5) = -10$

The last pair gives us what we need. We put those values into the binomials to get our final answer.

$$
\begin{aligned}
x^2 - 10x + 25 &= (x-5)(x-5) \\
&= (x-5)^2
\end{aligned}
$$

Check:

$$
\begin{aligned}
(x-5)^2 &= (x-5)(x-5) \\
&= x(x-5) - 5(x-5) \\
&= x^2 - 5x - 5x + 25 \\
&= x^2 - 10x + 25 \qquad \text{Confirmed!}
\end{aligned}
$$

Notice that this trinomial factors as the square of a binomial (in other words, a binomial times itself). Trinomials that factor into square of a binomial are called **Perfect Square Trinomials.**

Problem 7 : | *You Try*

Factor the following trinomials completely, if possible. If not factorable, identify it as prime.

a) $x^2 - 12x + 36$ b) $h^2 + 2h + 1$ c) $g^2 - 5g + 4$

Problem 8 : | *Worked Example*

Factor $x^2 - 16$ completely.

> **Solution.**

You might notice that this is not a trinomial. It is missing the linear term. However, we can "pencil in" the middle term with a coefficient of 0. (Remember that adding 0 to an expression does not change its value).

Consider then $x^2 + 0x - 16$. We need factors of -16 that sum to 0. Those factors are 4 and -4. Therefore,

$$x^2 - 16 = (x + 4)(x - 4)$$

Check:

$$
\begin{aligned}
(x + 4)(x - 4) &= x(x - 4) + 4(x - 4) \\
&= x^2 - 4x + 4x - 16 \\
&= x^2 - 16 \qquad \text{Confirmed!}
\end{aligned}
$$

Note, we can only find factors of c that sum to 0 when c is a negative perfect square. Consequently, when a binomial is the **difference of two squares**, it factors as: $a^2 - b^2 = (a + b)(a - b)$.

Problem 9 : | *You Try*

Factor the following binomials completely, if possible. If not factorable, identify it as prime.

a) $x^2 - 36$ c) $y^2 - 81$

b) $w^2 + 1$ d) $x^2 - 3$

When factoring any polynomial, you always want to first factor out the Greatest Common Factor (GCF).

Problem 10 : | Worked Example

Factor $5x^2 - 25x - 30$ completely.

Solution.

Factor out the GCF first: $5x^2 - 25x - 30 = 5(x^2 - 5x - 6)$.
Now we try to factor the trinomial inside the parenthesis. We want to find factors of -6 that sum to -5. The factors that work are -6 and 1.

$$5x^2 - 25x - 30 = 5(x^2 - 5x - 6)$$
$$= 5(x - 6)(x + 1)$$

Check:

$$5(x - 6)(x + 1) = (5x - 30)(x + 1)$$
$$= 5x(x + 1) - 30(x + 1)$$
$$= 5x^2 + 5x - 30x - 30$$
$$= 5x^2 - 25x - 30 \qquad \text{Confirmed!}$$

Problem 11 : | Worked Example

Factor the following trinomials completely, if possible. If not factorable, identify it as prime.

a) $3y^2 + 24y + 48$ b) $2x^3 - 10x^2 - 12x$ c) $5m^2 - 20$

Problem 12 : | *You Try*

Factor the following trinomials completely, if possible. If not factorable, identify it as prime.

a) $2p^2 - 2$

c) $4x^2 + 8x - 32$

b) $y^3 - 6y^2 + 9y$

d) $10c^3 + 60c^2 + 10c$

Problem 13 : | *Class/Media Example*

Given the quadratic and constant terms of a trinomial, what are all the possible options for the linear term, positive and negative: $x^2 + \_\_ + 12$

Problem 14 : | *You Try*

a) Given the quadratic and constant terms of a trinomial, what are all the possible options for the linear term, positive and negative: $x^2 + \_\_ + 30$

b) Given the quadratic and linear terms of a trinomial, what are all the possible positive options for the constant term: $x^2 + 7x + \_\_$

c) Given the quadratic and linear terms of a trinomial, what are all the possible positive options for the constant term: $x^2 - 8x + \_\_$

Practice Problems: *Factoring $ax^2 + bx + c$ - Part I*

Factor each expression completely.

1. $p^2 + 17p + 72$

2. $x^2 + x - 72$

3. $n^2 - 9n + 8$

4. $x^2 + x - 30$

5. $x^2 - 9x - 10$

6. $b^2 + 12b + 32$

7. $y^2 + 3y + 4$

8. $b^2 - 17b + 70$

9. $x^2 + 3x - 70$

10. $x^2 + 3x - 18$

11. $n^2 - 8n + 15$

12. $a^2 - 6a - 27$

13. $p^2 + 15p + 54$

14. $p^2 + 7p - 30$

15. $n^2 - 15n + 56$

16. $x^2 - 9$

17. $m^2 - 49$

18. $p^2 - 1$

19. $x^2 - 2x + 1$

20. $m^2 + 12x + 36$

21. $y^2 + 16y + 64$

22. $n^2 + 2n + 4$

23. $3v^2 - 12v + 18$

24. $-x^2 + 18x - 81$

25. $6x^2 + 18x + 12$

26. $4x^2 + 20x + 24$

27. $-5n^2 + 45n - 40$

28. $5v^3 + 20v^2 - 25v$

29. $4n^2 - 64$

30. $y^3 - 25y$

6.3. FACTORING $ax^2 + bx + c$ – Part II

Objective: To factor trinomials where the coefficient of the squared term is not equal to one.

Since factoring is the reverse process of multiplication, let's review what happens when you multiply two binomials like $(2x+3)(3x+1)$

$$\begin{aligned}(2x+3)(3x+1) &= 6x^2+2x+9x+3 \\ &= 6x^2+11x+3\end{aligned}$$

We notice that

- We obtained the quadratic term, $6x^2$, when we multiplied the leading terms of our binomials: $2x \cdot 3x = 6x^2$

- We obtained the constant term, 3, when we multiplied the second terms of our binomials: $3 \cdot 1 = 3$

- We obtained the linear term, $11x$, in a more complicated way: it is the sum of the two terms: $2x + 9x = 11x$

The strategy we will use to factor a trinomial will focus on the first and last terms of the trinomial. This could give us several options to check.

Problem 1 :	*Worked Example*

Factor $3x^2 + 7x + 2$

Solution.

Notice that

- to obtain the quadratic term, $3x^2$, we have only one possiblity: $3x \cdot x$. So the binomials we are looking for should have the form $(3x + \underline{})(x + \underline{})$.

- to obtain the constant term, 2, we have only one possibility: $2 \cdot 1$. So the binomials we are looking for have the form $(\underline{} + 2)(\underline{} + 1)$.

This gives a total of 2 possibilities for the binomials:
$$(3x+2)(x+1) \quad \text{or} \quad (3x+1)(x+2)$$
To check which one is correct we multiply

$$\begin{aligned}(3x+2)(x+1) &= 3x^2+3x+2x+2 \\ &= 3x^2+5x+2\end{aligned} \qquad \begin{aligned}(3x+1)(x+2) &= 3x^2+6x+x+2 \\ &= 3x^2+7x+2\end{aligned}$$

So the second option is the correct factorization.

The method described above is known as the *list and check* method (or *guess and check method)*. With practice we will become more efficient with this method, at least when the numbers involved are simple enough.

Problem 2 : | *You try*

Factor $2x^2 + 11x + 5$

We should note that these first examples were simple because the coefficients of both the quadratic term and the constant term were prime numbers. This will not always be the case.

Problem 3 : | *Worked Example*

Factor $6x^2 + 7x + 2$

Solution.

Solution: Notice that

- to obtain the quadratic term $6x^2$ we have two options: $6x \cdot x$ and $3x \cdot 2x$. So the binomials we are looking for should have the form
  $(6x + \underline{})(x + \underline{})$ or $(3x + \underline{})(2x + \underline{})$.

- to obtain the consant term we have only one option: $2 \cdot 1$

This means we have a total of 4 possibilities for our binomials:

a) $(6x + 1)(x + 2)$ b) $(6x + 2)(x + 1)$

c) $(3x + 1)(2x + 2)$ d) $(3x + 2)(2x + 1)$

We can check each one of these possibilities and see that the correct option is d), since

$$(3x + 2)(2x + 1) = 6x^2 + 3x + 4x + 2$$
$$= 6x^2 + 7x + 2$$

Problem 4 : | *Class/Media Example*

Factor the polynomial $3x^2 - 14x + 8$

Problem 5 : | *You Try*

Factor the polynomial $3x^2 + 14x + 8$

Problem 6 : | *Worked Example*

Factor the trinomial $2x^2 - 7x + 6$

Solution.

- Notice that since the linear term is negative and the constant term is positive, the pair of binomials we are looking for must have the form $(\_\_ - \_\_)(\_\_ - \_\_)$.

- To obtain the qudaratic term $2x^2$ we have only one option: $2x \cdot x$. So the binomials must have the form $(2x - \_\_)(x - \_\_)$.

- To obtain the linear term 6 we have two options: $6 \cdot 1$ and $3 \cdot 2$. So the binomials can have the form $(\_\_ - 6)(\_\_ - 1)$ or $(\_\_ - 3)(\_\_ - 2)$

We now create a list of all possible options:

a) $(2x - 6)(x - 1)$ b) $(2x - 1)(x - 6)$
c) $(2x - 3)(x - 2)$ d) $(2x - 2)(x - 3)$

Notice that we can rule out a) and d):

- the binomial $(2x - 6)$ in a) has a common factor of 2. If this was one of our binomials, then there would be a common factor of 2 in our original trinomial, but there is not.

- the binomial $(2x - 2)$ in d) has a common factor of 2 as well. If this was one of our binomials, then there would be a common factor of 2 in our original trinomial, but there is not.

So there are only two cases to check. After performing the multiplications we conclude that the correct factorization is c).

$$(2x-3)(x-2) = 2x^2-4x-3x+6$$
$$= 2x^2-7x+6$$

We can summarize what happened in the last example as follows:

If the original trinomial does not have a common factor, then there can be no common factor in any of the binomials

Problem 7 : | *Worked Example*

Factor the polynonial $15x^2+26x+8$

Solution.

- To obtain the quadratic term $15x^2$ we have two options: $15x \cdot x$ and $3x \cdot 5x$. So the binomials can have the form $(15x+\_)(x+\_)$ or $(3x+\_)(5x+\_)$.

- To obtain the linear term 8, we have two options: $8 \cdot 1$ and $4 \cdot 2$. So the binomials can have the form $(\_+8)(\_+1)$ or $(\_+4)(\_+2)$

This give us a total of 8 combinations! Here is the list:
a) $(15x+8)(x+1)$ b) $(15x+1)(x+8)$ c) $(15x+4)(x+2)$ d) $(15x+2)(x+4)$
e) $(3x+8)(5x+1)$ f) $(3x+1)(5x+8)$ g) $(3x+4)(5x+2)$ h) $(3x+2)(5x+4)$

Instead of checking all 8 possibilities, we can notice that since all the terms are positive and the middle term of our trinomial is *not so big* ($26x$), we can rule out cases b), c) d) and e), since these will create a *large* middle term ($15x \cdot 8 = 120x$, $15x \cdot 2 = 30x$, $15x \cdot 4 = 60x$ and $8 \cdot 5x = 40x$).

This leaves us with 4 cases to check. After performing all the multiplications we notice that the correct factorization is g) since

$$(3x+4)(5x+2) = 15x^2+6x+20x+8$$
$$= 15x^2+26x+8$$

The number of cases to handle in the last example may be getting too big for this method, and some people prefer to have a different method. The method we desscribe below is based on reversing the steps of the multiplication. Let's multiply working out each step of the distributive property:

$$(3x+4)(5x+2) = 3x(5x+2)+4(5x+2) \quad \text{step (1)}$$
$$= 15x^2+6x+20x+8 \quad \text{step (2)}$$
$$= 15x^2+26x+8 \quad \text{step (3)}$$

Notice that

- if we multiply the coefficient of the quadrataic term and the coeficient of the linear term of our trinomial in step (3) we get $15 \cdot 8 = 120$.

- if we multiply the two coefficients of the linear terms in step (2) we also get 120: $6 \cdot 20 = 120$

This observation will be true always and we can use this fact to reverse the steps.
In order to do this we must understand how to

A) re-write the linear term of the trinomial as the sum of two appropriate terms (go from step (3) to step (2))

B) factor the gcf from the first two terms and the gcf from the last two terms (go from step (2) to step (1)).

C) use $(5x + 2)$ as a common factor (go from the left side of step (1) to the right side of step (1)).

Steps **B)** and **C)** should be familar: they are the instructions for *factoring by gouping* the 4-terms obtained in step **A)**.

Notice that a key part of this process is that we needed to find a pair of numbers that will multiply 120, and that add up to 26. In practice this might take some time, since we might have to list all possible pairs of factors of 120 until we find a pair that adds up to 26.

This method for factoring a trinomial of the form $ax^2 + bx + c$ is known as the **ac-method**, since the first step will be to multiply the coefficient of the quadratic term, a, and the constant term, c, of the trinomial.

Problem 8 : | *Worked Example*

Factor the trinomial $6x^2 + 19x + 15$

Solution.

Although the number of cases in this problem is still easy to handle using the list-and-check method, we will use the ac-method in order to illustrate the steps.

- multiply $ac = 6 \cdot 15 = 90$

- we need to find a pair of numbers that multiply 90, and with a sum of 19. The numbers are easily found to be 9 and 10.

- we are ready to reverse the steps:

$$6x^2+19x+15 \;=\; 6x^2+9x+10x+15$$ use the pair of numbers we found to rewrite the linear term of the trinomial

$$=\; 3x(2x+3)+5(2x+3)$$ find the gcf of the first two terms, and the gcf of the last two terms

$$=\; (3x+5)(2x+3)$$ use $(2x+3)$ as a common factor

Problem 9 : | *Worked Example*

Factor the trinomial $18x^2+45x+25$

  Solution.

If we want to use the list and check method the number of cases needed to for this problem may get too big. We will use the *ac*-method.

– multiply $ac=15\cdot30=450$

– we need to find a pair of numbers that multiply 450 and add up 45. We can list all possible ways to factor 450:
    1(450)   2(225)   3(150)   5(90)
    6(75)   9(50)   10(45)   15(30)   18(25)

– The pair that we are looking for is 15 and 30. We can now work out the steps.

$$18x^2+45x+25 \;=\; 18x^2+30x+15x+25$$ use the two numbers found above to re-write the linear term

$$=\; 6x(3x+5)+5(3x+5)$$ factor the gcf of the first two terms and the gcf of the last two terms

$$=\; (6x+5)(3x+5)$$ use $(3x+5)$ as a common factor

One important observation is that the order in which we write the two terms in our first step does not matter. To illustrate this we show the work for the last example changing the order:

$$18x^2+45x+25 \;=\; 18x^2+15x+30x+25$$ re-writing the two terms in reverse order

$$=\; 3x(6x+5)+5(6x+5)$$ factor the gcf of the first two terms and the gcf of the last two terms

$$=\; (3x+5)(6x+5)$$ use $(6x+5)$ as a common factor

Problem 10 : | *You Try*

Factor the trinomial $8x^2+37x+20$

Problem 11 : | *Worked Example*

Factor the trinomial $6x^2+19x-36$

Solution.

Using the *ac*-method

- $ac=6(-36)=-216$

- factors of 216 (we will take care of the sign later)
  1 (216)   2(108)   3(72)   4(54)
  6(36)   8(27)   9(24)   12(18)

- since the product *ac* was negative (−216), the pair of numbers we are looking for must have opposite signs, which means that we are looking for a pair in the above list with a **difference** of 21. The pair is (-8) and 27.

- We can now work out the factoring:

$$6x^2+19x-36 = 6x^2+27x-8x-36 \qquad \text{re-write the linear term}$$

$$= 3x(2x+9)-4(2x+9) \qquad \text{factor the corresponding gcf's}$$

$$= (3x-4)(2x+9) \qquad \text{our factorization.}$$

Problem 12 : | *Class/Media Example*

Use any method to factor the trinomial $12x^2 - 7x - 10$.

Problem 13 : | *You Try*

Use any method to factor the trinomial $6x^2 + 19x + 20$.

One useful application of the ac-method is that it gives us a tool to determine if a trinomial is prime.

Problem 14 : | *Worked Example*

Factor the trinomial $6x^2 + 25x + 30$

    Solution.

    – $ac = 180$

    –   list of factors of 180:
         1(180)   2(90)   3(60)   4(45)
         5(36)   6(30)   9(20)   10(18)   12(15)

Since none of these pairs add up to 5, we can conclude that the trinomial $6x^2 + 25x + 30$ is prime.

Problem 15 : | *Class/Media Example*

Factor the trinomial if possible. If not, state why is it prime.
$12x^2+9x-10$

Problem 16 : | *Class/Media Example*

Factor the trinomial if possible. If not, state why is it prime.
$8x^2-12x-10$

Problem 17 : | *Worked Example*

Factor the trinomial if possible. If not, state why is it prime.
$9x^2+24x+16$

Solution.

We will use the ac-method.

– $ac=9\cdot16=144$

– We are looking for two numbers that multiply 144 and add up 24. We don't need a list in this case, the numbers are easily found to be 12 and 12.

– We now work out the factoring:

$$9x^2 + 24x + 16 = 9x^2 + 12x + 12x + 16 \quad \text{re-write the linear term}$$

$$= 3x(3x+4) + 4(3x+4) \quad \text{factor the corresponding gcf's}$$

$$= (3x+4)(3x+4) \quad \text{our factorization.}$$

$$= (3x+4)^2$$

Notice that in the last example we could have identified our trinomial as a perfect square and find the factorization direclty.

Practice Problems: *Factoring $ax^2 + bx + c$ Part II*

Factor the trinomial if possible. If not, state why is it prime.

1. $4x^2 + 4x - 3$

2. $5x^2 + 32x + 12$

3. $3x^2 + 4x - 8$

4. $6x^2 + 7x - 3$

5. $7x^2 - 11x - 6$

6. $9x^2 - 6x - 8$

7. $8x^2 + 2x - 3$

8. $10x^2 + 9x + 2$

9. $15x^2 + 14x + 3$

10. $2x^2 + 25x + 12$

11. $12x^2 - 2x - 4$

12. $21x^2 - 5x - 6$

13. $28x^2 - 13x - 6$

14. $8x^2 + 14x - 15$

15. $20x^2 + 7x - 6$

6.4. FACTORING SPECIAL PRODUCTS

Objective: Factor the difference of two squares and recognize perfect square trinomials.

Difference of Two Squares

In this section, we learn how to recognize the pattern for the difference of two squares, $a^2 - b^2$. If we see the pattern that occurs for the difference of two squares, we can quickly factor these terms.

Since factoring is the reverse process of multiplication, let us review the product of the polynomial operation, $(a+b)(a-b)$.

$$\begin{aligned} (a+b)(a-b) &= a(a-b)+b(a-b) \\ &= a^2-ab+ab-b^2 \\ &= a^2-b^2 \end{aligned}$$

We notice that the middle terms are opposites of each other and when combined, will zero out, leaving us with the square of the first term minus the square of the last term. Recognizing this pattern will help us easily factor the difference of two squares. Two binomials will form that are identical except for their signs. One binomial will have a positive sign; the other will have a negative sign. In the pattern, a will be the square root of the first term and b will be the square root of the second term.

> **Difference of Two Squares:**
> $$a^2 - b^2 = (a+b)(a-b)$$

Problem 1 : *Worked Example*

Factor $m^2 - 25$ completely.

   Solution.

$$m^2 - 25 = (m)^2 - (5)^2 \qquad \text{Find square root of first \& second binomial term}$$

$$= (\_\_ + \_\_)(\_\_ - \_\_) \qquad \text{Form two sets of parenthesis with opposite signs}$$

$$= (m+5)(m-5)$$

Check:

$$\begin{aligned} (m+5)(m-5) &= m(m-5)+5(m-5) \\ &= m^2-5m+5m-25 \\ &= m^2-25 \qquad \text{Confirmed!} \end{aligned}$$

Warning.
It is important to note that the sum of two squares will not factor unless it contains a greatest common factor.

Sum of Squares: $a^2 + b^2$ is Prime

Problem 2 :	*Worked Example*

Factor $x^2 + 16$ completely. If binomial is unfactorable, identify it as prime.

  Solution.

There are no factors that when multiplied together give the product $x^2 + 16$. A common mistake is to write $(x+4)(x+4)$ as the factors. If we multiply these two binomials, we see that the product is not the original binomial.

$$\begin{aligned}
(x+4)(x+4) &= x(x+4)+4(x+4) \\
&= x^2 + 4x + 4x + 16 \\
&= x^2 + 8x + 16 \\
&\neq x^2 + 16
\end{aligned}$$

Therefore, $x^2 + 16$ is not factorble and is prime.

Problem 3 :	*Media/Class Example*

Factor the following binomials completely. If binomial is unfactorable, identify it as prime.

  a) $x^2 - 16$ b) $1 - 36p^2$

  c) $y^2 + 49$ d) $3m^2 - 27$

  e) $x^4 - 16$ f) $36 + p^2$

Problem 4 : | *You Try*

Factor the following binomials completely. If binomial is unfactorable, identify it as prime.

a) $25m^2 - 9$

d) $1 - 49p^2$

b) $h^2 + 36$

e) $2c^2 - 32$

c) $x^4 - 81$

f) $36 - 4x^2$

Perfect Square Trinomials

In previous sections, we practiced factoring trinomials in the form $ax^2 + bx + c$. Trinomial such as $x^2 + 10x + 25$, has a special pattern and factors to a perfect square. The factors of this expression are two identical binomials $(x + 5)(x + 5)$. We can rewrite these factors using exponential notation $(x + 5)^2$ and notice that they form a perfect square.

It is not always easy to see this pattern. We can always use our previous methods for factoring trinomials. However, if we recognize the perfect square pattern, we can eliminate some steps and save time.

Recall from our work on multiplication of polynomials that:
$$(a + b)(a + b) = a^2 + 2ab + b^2 \qquad \text{and} \qquad (a - b)(a - b) = a^2 - 2ab + b^2$$
To recognize this pattern, first determine if the sign pattern in the binomial will be the same. The pattern in the original trinomial must be either

$$(\_\_ + \_\_ + \_\_) \quad \text{or} \quad (\_\_ - \_\_ + \_\_)$$

Next, notice if the quadratic term and the constant term are perfect squares. If the quadratic term and the constant term are perfect squares, and the product of the square root of each of these terms multiplied by 2 gives the linear term of the trinomial, then you have a pattern for a perfect square trinomial.

> **Perfect Square Trinomial:**
> $$a^2 + 2ab + b^2 = (a+b)^2 \text{ or}$$
> $$a^2 - 2ab + b^2 = (a-b)^2$$

Problem 5 : | *Media/Class Example*

Factor $x^2 - 6x + 9$ completely.

Solution.

Notice that the quadratic term, x^2, and the constant term, 9, are perfect squares. The square root of the quadratic term is x and the square root of the constant term is 3. When multiplied together, the product is $3x$. When $3x$ is multiplied by 2, we get the linear term, $6x$, which is the linear term of the trinomial. This indicates that the original expression is a perfect square trinomial.

$$\begin{aligned} x^2 - 6x + 9 &= (x)^2 - 2(x)(3) + (3)^2 \\ &= (x-3)(x-3) \\ &= (x-3)^2 \end{aligned}$$

Check:

$$\begin{aligned} (x-3)^2 &= (x-3)(x-3) \\ &= x(x-3) - 3(x-3) \\ &= x^2 - 3x - 3x + 9 \\ &= x^2 - 6x + 9 \qquad\qquad \text{Confirmed!} \end{aligned}$$

Problem 6 : | *Worked Example*

Factor $25x^2 + 10x + 1$ completely.

Solution.

Notice that the quadratic term, $25x^2$, and the constant term, 1, are perfect squares. The square root of the quadratic term is $5x$ and the square root of the constant term is 1. When multiplied together, the product is $5x$. When $5x$ is multiplied by 2, we get the linear term, $10x$, which is the linear term of the trinomial. This indicates that the original expression is a perfect square trinomial.

$$\begin{aligned} 25x^2 + 10x + 1 &= (5x)^2 - 2(5x)(1) + (1)^2 \\ &= (5x-1)(5x-1) \\ &= (5x-1)^2 \end{aligned}$$

Check.

$$
\begin{aligned}
(5x-1)^2 &= (5x-1)(5x-1) \\
&= 5x(5x-1)-1(5x-1) \\
&= 25x^2-5x-5x+1 \\
&= 25x^2-10x+1 \qquad \text{Confirmed!}
\end{aligned}
$$

Problem 7 : | *Media/Class Example*

Factor the following trinomials completely.

a) $m^2-12m+36$

b) $4x^2+20x+25$

Problem 8 : | *You Try*

Factor the following trinomials completely.

a) $y^2+8y+16$

c) $25-30k+9k^2$

b) $49p^2-28p+4$

d) $9x^2+12x+4$

Problem 9 : | *Worked Example*

Factor $48m^2 - 24m + 3$ completely.

> **Solution.**

First find GCF, if there is one. The GCF is 3.

$$48m^2 - 24m + 3 = 3(16m^2 - 8m + 1)$$

Next, work on the trinomial, $16m^2 - 18m + 1$. Notice that the quadratic term, $16m^2$, and the constant term, 1, are perfect squares. The square root of the quadratic term is $4m$ and the square root of the constant term is 1. When multiplied together, the product is $4m$. When $4m$ is multiplied by 2, we get the linear term, $8m$, which is the linear term of the trinomial. This indicates that the original expression is a perfect square trinomial.

$$
\begin{aligned}
48m^2 - 24m + 3 &= 3(16m^2 - 8m + 1) \\
&= 3[(4m)^2 - 2(4m)(1) + (1)^2] \\
&= 3[(4m - 1)(4m - 1] \\
&= 3(4m - 1)^2
\end{aligned}
$$

Check.

$$
\begin{aligned}
3(4m - 1)^2 &= 3(4m - 1)(4m - 1) \\
&= 3[4m(4m - 1) - 1(4m - 1] \\
&= 3[16m^2 - 4m - 4m + 1] \\
&= 3[16m^2 - 8m + 1] \\
&= 48m^2 - 24m + 3 \qquad \text{Confirmed!}
\end{aligned}
$$

Problem 10 : | *Media/Class Example*

Factor $50n^2 - 40n + 8$ completely.

Problem 11 : | *You Try*

Factor the following trinomials completely.

a) $36x^2 + 24x + 4$ b) $5m^2 - 40m + 80$

The following table summarizes factoring special products.

Factoring Special Products

Difference of Squares	$a^2 - b^2$	$=$	$(a+b)(a-b)$
Sum of Squares	$a^2 + b^2$		Prime
Perfect Square Trinomial	$a^2 + 2ab + b^2$	$=$	$(a+b)^2$
Perfect Square Trinomial	$a^2 - 2ab + b^2$	$=$	$(a-b)^2$

As always, when factoring, it is important to check for GCF first. Only after checking for GCF should we be using the special products.

Practice Problems: *Factoring Special Products*

Factor each trinomial completely, if possible. If the trinomial is not factorable, identify it as prime.

1. $x^2 - 9$

2. $x^2 - 1$

3. $v^2 + 25$

4. $4 - p^2$

5. $4v^2 - 1$

6. $9k^2 - 4$

7. $1 - 9a^2$

8. $3x^2 - 27$

9. $125x^2 + 45$

10. $5n^2 - 20$

11. $18a^2 - 50$

12. $64 + 4m^2$

13. $a^2 - 2a + 1$

14. $k^2 + 4k + 4$

15. $n^2 - 8n + 16$

16. $25p^2 - 10p + 1$

17. $4k^2 + 28k + 49$

18. $x^2 + 8x + 16$

19. $25a^2 + 30a + 9$

20. $4a^2 - 20a + 25$

21. $18m^2 - 24m + 8$

22. $5x^2 + 10x + 5$

23. $20x^2 + 20x + 5$

24. $8x^2 - 24x + 18$

25. $a^4 - 81$

26. $n^4 - 1$

27. $16 - z^4$

Rescue Roody!

Roody was told to factor $16x^2 - 36$ completely. Roody recognized that the binomial is a difference of two squares. This is how Roody factored the binomial.

$$16x^2 - 36 = (4x + 6)(4x - 6)$$

Roody then checked his work.

$$
\begin{aligned}
(4x + 6)(4x - 6) &= 4x(4x - 6) + 6(4x - 6) \\
&= 16x^2 - 24x + 24x - 36 \\
&= 16x^2 - 36
\end{aligned}
$$

Everything seems fine but his answer was marked as incorrect. Help Roody.

6.5. FACTORING STRATEGIES

Objective: Idenfity and use the correct method to factor various polynomials.

With so many different methods to factor, it is easy to get lost as to which method to use and when. We will organize the different factoring strategies we have seen.

Factoring Strategy:

1. Always factor greatest common factor first!

2. Look at the number of terms in the polynomial.

    a) 2 Terms:

        i. Sum of Squares: $a^2 + b^2$ is prime

        ii. Difference of Squares: $a^2 - b^2 = (a+b)(a-b)$

    b) 3 Terms:

        i. Use the strategy in the previous sections to factor trinomial of the form $ax^2 + bx + c$

        ii. Perfect Square Trinomial:

            • $a^2 + 2ab + b^2 = (a+b)^2$

            • $a^2 - 2ab + b^2 = (a-b)^2$

    c) 4 Terms: Group

It is important to be comfortable and confident with using factoring methods and deciding which method to use.

Problem 1 : | *Worked Example*

Factor $7ax - 14x + 3a - 6$ completely.

    Solution.

$$7ax - 14x + 3a - 6 \qquad \text{No GCF; 4 terms, try grouping}$$

$$= \underbrace{7ax - 14x} + \underbrace{3a - 6} \qquad \text{Group first 2 terms and last 2 terms}$$

$$= 7x(a-2) + 3(a-2) \qquad \text{Find GCF for each group and factor}$$

$$= (a-2)(7x+3) \qquad \text{Factor out GCF, } (a-2)$$

Check to verify that the factors are correct.

$$(a-2)(7x+3) = a(7x+3)-2(7x+3)$$
$$= 7ax+3a-14x-6$$
$$= 7ax-14x+3a-6 \qquad \text{It checks!}$$

Problem 2 : | *Worked Example*

Factor $a^2-22a-48$ completely.

   Solution.

Trinomial has no greatest common factor. We need to find factors of -48 that sum up to -22. Given that the product is negative, we have to consider combinations of positive and negative factors.

Factors of -48	Product of Factors	Sum of Factors
$1,-48$	$1 \cdot (-48) = -48$	$1+(-48) = -47$
$2,-24$	$2 \cdot (-24) = -48$	$2+(-24) = -22$

We can stop at this point since we have found the pair we need. We put those values into the binomials to get our final answer.

$$a^2-22a-48 = (a+2)(a-24)$$

Check our factorization.

$$(a+2)(a-24) = a(a-24)+2(a-24)$$
$$= a^2-24a+2a-48$$
$$= a^2-22a-48 \qquad \text{It checks!}$$

Problem 3 : | *Worked Example*

Factor $100y^2-400$ completely.

   Solution.

$$100y^2-400 \qquad \text{Factor GCF first}$$
$$= 100(y^2-4) \qquad \text{Two terms, difference of squares}$$
$$= 100(y+2)(y-2) \qquad \text{Factoring completed!}$$

Check to verify factors are correct.

$$100(y+2)(y-2) = 100[y(y-2)+2(y-2)]$$
$$= 100[y^2-2y+2y-4]$$
$$= 100[y^2-4]$$
$$= 100y^2-400 \qquad \text{It checks!}$$

Problem 4 : | *Worked Example*

Factor $6n^3 + 14n^2 - 2n$ completely.

Solution.

$$6n^3 + 14n^2 - 2n \qquad \text{Factor GCF first}$$
$$= 2n(3n^2 + 7n - 1) \qquad \text{Factor trinomial}$$
$$= 2n(3n^2 + 7n - 1) \qquad \text{Trinomial is prime; factoring completed!}$$

Check to verify factors are correct.

$$2n(3n^2 + 7n - 1) = 6n^3 + 14n^2 - 2n \qquad \text{It checks!}$$

Problem 5 : | *Media/Class Example*

Factor the following expressions completely. If unfactorable, identify as prime.

a) $4x^2 + 56x + 196$

c) $5 - 45m^2$

b) $xy + 3y - 7x - 21$

d) $36p^2 - 13p^2 + 1$

Problem 6 : | *You Try*

Factor the following expressions completely. If unfactorable, identify as prime.

a) $3m^2 - 3m - 18$

c) $15p^2 + p - 2$

b) $4n^2 - 8n - 4$

d) $2x^3 + x^2 - 8x - 4$

Practice Problems: *Factoring Strategies*

Factor each expression completely. If not factorable, identify as prime.

1. $2p(3p+4)+(3p+4)$

2. x^2+4x+3

3. m^2-4

4. v^3+v

5. n^2-3n+4

6. $2x^2-11x+15$

7. $2x^2-10x+12$

8. $ab-3a+2b-6$

9. n^3-5n^2-6n

10. $x^2+8x+16$

11. $5x^2-22x-15$

12. n^3+7n^2+10n

13. $8y^3+6y^2+20y+15$

14. $16a^2-9$

15. $5n^2+7n-6$

16. $16x^2-72x+81$

17. $2k^2+k-10$

18. $9n^3-6n^2+3n$

19. $2x^3+6x^2-20x$

20. $3xy-15y-x+5$

21. $9-25y^2$

22. $4x^2+24xy+36y^2$

23. $3k^3-27k^2+60k$

24. $5u^2-9u+4$

25. $45m^2-150m+125$

26. $27m^2-48$

27. x^4+4x^2

28. $n-n^3$

29. p^4-81

30. $y^3+5y^2-4y-20$

6.6. SOLVING QUADRATIC EQUATIONS BY FACTORING

Objective: To solve factoring techniques to solve quadratic equations.

We begin this section by stating a basic fact about real numbers: *if the product of two numbers is zero, then at least one of the numbers must be equal to zero.* This is known as the zero-product property, and can be written as

Zero Product Property
If $a \cdot b = 0$
then either
$a = 0$ or $b = 0$

Problem 1 : | *Worked Example*

Use the zero-product property to solve the equation $(x+1)(x+2) = 0$.

> **Solution.**

The zero-product property tells that if $(x+1)(x+2) = 0$ then we must have
$$x + 1 = 0 \quad \text{or} \quad x + 2 = 0$$
We now have two (linear) equations to solve, and each one will give us a solution to our orignal problem.

$$
\begin{array}{ll}
x + 1 = 0 & x + 2 = 0 \\
\underline{\quad -1 \quad -1} & \underline{\quad -2 \quad -2} \\
x = -1 & x = -2
\end{array}
$$

We can easily verify our solutions using these values in the original equation:

- using $x = -1$, we obtain
  $0\,(1) = 0$ ✓

- using $x = -2$ we obtain
  $(-1)\,0 = 0$ ✓

On our previous example the original equation was given in factored form. Usually this will not be the case.

Problem 2 : | *Worked Example*

Solve the equation $2x^2 + 4x = 0$. Check your solutions.

> **Solution.**

We must factor first.
$$
\begin{array}{ll}
2x^2 + 4x = 0 & \\
2x(x+2) = 0 & \text{factor the GCF}
\end{array}
$$

Using the zero-product property we must have either $2x = 0$ or $x + 2 = 0$. We solve each equation separately

$$2x \ = \ 0 \quad \text{or} \quad x + 2 \ = \ 0$$
$$\underline{\quad -2 \ = \ -2}$$
$$\frac{2x}{2} \ = \ \frac{0}{2} \qquad\qquad x \ = \ -2$$

$$x \ = \ 0 \quad \text{or} \quad x = -2$$

Let's check our solutions.

- using $x = 0$
  $0^2 + 0 = 0$ ✓

- using $x = -1$
  $(-1)^2 + (-1) \ = \ 1 - 1$
  $\qquad\qquad\qquad = \ 0$ ✓

Problem 3 : | *Class/Media Example*

Solve the equation $3x^2 - 6x = 0$. Check your solutions.

Problem 4 : | *You Try*

Solve the equation the following equations. Check your solutions.

a) $(3x - 4)(x + 5) = 0$

b) $4x^2 - 8x = 0$

Problem 5 : | *Worked Example*

Solve the equation $x^2 + 5x - 6 = 0$

Solution.

$$\begin{aligned} x^2 + 5x - 6 &= 0 \\ (x+6)(x-1) &= 0 \end{aligned}$$

either $x+6=0$ or $x-1=0$

we now solve each: $x+6=0$ $x-1=0$

$x=-6$ or $x=1$

Problem 6 : | *Class/Media Example*

Solve the equation $x^2 - 5x - 14 = 0$.

Problem 7 : | *You Try*

Solve the equation $x^2 + x - 12 = 0$

Notice that to use the zero-product property we need to have a zero on one side of the equation, and a non-prime polynomial on the other side. If that is not the case we must first re-write the equation to have it in this form.

Problem 8 : | *Worked Example*

Solve the equation $x^2 + 7x = 18$. Check your answers.

    Solution.

In order to use the zero-product property we have to have a zero on one side of the equation.

$$x^2 + 7x = 18$$
$$\underline{\quad -18 \quad -18} \quad \text{subtract 18 from each side}$$
$$x^2 + 7x - 18 = 0$$

We can now factor and use the zero-product property.

$$x^2 + 7x - 18 = 0$$
$$(x + 9)(x - 2) = 0$$

$$\text{either } x + 9 = 0 \qquad \text{or} \qquad x - 2 = 0$$

$$\text{we now solve each: } x + 9 = 0 \qquad\qquad x - 2 = 0$$

$$x = -9 \qquad \text{or} \qquad x = 2$$

Problem 9 : | *Class/Media Example*

Solve the following equations.

a) $x^2 + 8x = 20$ b) $2x^2 + 3x = 12 - 2x$

Problem 10 : | *You Try*

Solve the following equations.

a) $x^2 + 4x = 21$ b) $3y^2 = 6y$

Problem 11 : | *Worked Example*

Solve the equation $x^2 - 25 = 0$

Solution.

Recall that we can factor the right side as a difference of squares. We can then use the zero-product property.

$$x^2 - 25 = 0$$
$$(x+5)(x-5) = 0$$

either $x+5 = 0 \qquad$ or $\qquad x-5 = 0$

we now solve each: $x+5 = 0 \qquad\qquad x-5 = 0$

$$x = -5 \quad \text{or} \quad x = 5$$

Problem 12 : | *Worked Example*

Solve the equation $x(x+1) = 12$

Solution.

Remember that you cannot use the zero product property unless the equation is set to zero and the polynomial is factored. Let us subtract 12 from each side of the equation first.

$$x(x+1) = 12$$
$$x(x+1) - 12 = 0$$
$$x^2 + x - 12 = 0$$
$$(x+4)(x-3) = 0$$

Use Zero Product Property:

$$x+4 = 0 \text{ or } x-3 = 0$$
$$x = -4 \text{ or } x = 3$$

Problem 13 : | *Class/Media Example*

Solve the following equations

a) $x^2 = 49$

b) $(x+2)^2 + x^2 = 10$

Problem 14 : | *You Try*

Solve the following equations.

a) $2x^2 = 8$ b) $x(3x - 5) = 2$

We can also use these ideas to go backwards, that is, given two numbers p and q, find a quadratic function with solutions p and q.

Problem 15 : | *Worked Example*

Give a quadratic equation with integer coefficients that has solutions $x = 3$ and $x = -5$.

   Solution.

We can assume that the equation will have the form $ax^2 + bx + c = 0$.
If we want $x = 3$ to be a solution of our equation, then $(x - 3)$ should be a factor of our polynomial.
If we want $x = -5$ to be a solution, then $(x + 5)$ should be a factor of our polynomial.
So our equation becomes

$$(x - 3)(x + 5) = 0$$

Which can be re-written as

$$x^2 + 2x - 15 = 0$$

Problem 16 : | *Class/Media Example*

Give a quadratic equation with integer coefficients that has solutions $x = 4$ and $x = -\dfrac{3}{2}$

Problem 17 : | *You Try*

Give a quadratic equation with integer coefficients that has solutions $x = -2$ and $x = \dfrac{3}{4}$

Practice Problems: *Solving Quadratic Equations by Factoring*

Solve each equation by factoring.

1. $x^2 + 5x + 6 = 0$

2. $x^2 - 6x = 0$

3. $p^2 + 3p = 0$

4. $t^2 = t$

5. $5x^2 + 6x = 0$

6. $6m^2 = 12m$

7. $x^2 + 4x + 3 = 0$

8. $m^2 - 2m = 8$

9. $p^2 - 10p + 24 = 0$

10. $n^2 - 6n = 0$

11. $2x^2 - 14x + 24 = 0$

12. $2x^2 = x + 6$

13. $a^2 + a = 6$

14. $n^2 - 15n + 36 = 0$

15. $x^2 - 23x + 130 = 0$

16. $5y^2 - 75y = -220$

17. $n^2 + 13n = -30$

18. $p^3 - p = 0$

19. $4x^2 + 4x + 1 = 0$

20. $n^2 + 7n + 6 = 0$

21. $2x^2 = 11x - 5$

22. $6k^2 - 17k = -5$

23. $9n^2 + 6n^2 = -1$

24. $x^3 + 6x^2 = 0$

25. $x^2 - 15x = 250$

Give a quadratic equation with integer coefficient that has the given solutions.

1. $x = 5$ and $x = 1$

2. $x = 3$ and $x = -5$

3. $x = -4$ and $x = 4$

4. $x = 7$ and $x = -3$

5. $x = 3$

6. $x = 4$

7. $x = \dfrac{2}{3}$ and $x = 4$

8. $x = \dfrac{3}{2}$ and $x = -\dfrac{1}{3}$

9. $x = 5$ and $x = \dfrac{1}{10}$

10. $x = -\dfrac{4}{5}$

CHAPTER 6 ASSESSMENT

Factor each expression completely.

1. $6x^3 + 14x^2$

2. $7c^3 - 14c^2 + 3c - 6$

3. $y^2 + 7y + 6$

4. $x^2 - 5x - 24$

5. $n^2 + 49$

6. $2y^2 + y - 3$

7. $4p^2 - 4p + 1$

8. $3c^2 + 24c + 48$

9. $1 - 25m^2$

10. $15a^2 - 10a - 25$

Solve each of the following equations by factoring.

11. $(3k+4)(k-5) = 0$

12. $x^2 + 12x + 20 = 0$

13. $p^2 - 4p = 0$

14. $m^2 + 9 = 6m$

15. $y(3y-1) = 10$

15. $3w^2 = 10w - 3$

17. $2y^2 = 8$

18. $4p^2 = 12p$

Chapter 7

Radicals Part I

7.1. Square Roots

Objective: Understand the meaning of square root; Differentiate between rational vs irrational numbers, exact values and approximations.

You may already know that the square root of 4 is 2 or that $\sqrt{25} = 5$. Let us formalize our understanding:

$$\sqrt{b} = a \text{ for any } b > 0 \text{ if } a^2 = b$$

In other words, $\sqrt{25} = 5$ because $5^2 = 25$.

Some important things to note:

- The $\sqrt{}$ symbol is known as the **radical**.

- The expression under the radical symbol is known as the **radicand**.

- $\sqrt{b}$ is known as the **principal root**, a positive value. $-\sqrt{b}$ gives us the negative square root.

In this seciton, we will only focus on problems where the radicands are constants. in a future chapter, we will see problems that include the square root of a variable.

Problem 1 : | *Worked Example*

Evaluate the following.

a) $\sqrt{16}$

b) $-\sqrt{81}$

Solution.

a) $\sqrt{16} = 4$ because $4^2 = 16$

b) $-\sqrt{81} = -9$. First evaluate $\sqrt{81} = 9$. We see that $-\sqrt{81}$ can be written as $(-1) \cdot \sqrt{81} = (-1) \cdot 9 = -9$

Problem 2 : | *Media/Class Example*

Evaluate the following.

a) $\sqrt{100}$

c) $-\sqrt{49}$

b) $\sqrt{1}$

d) $\sqrt{0.25}$

Problem 3 : | *You Try*

Evaluate the following.

a) $\sqrt{64}$

c) $-\sqrt{400}$

b) $\sqrt{0}$

d) $\sqrt{\dfrac{9}{4}}$

Problem 4 : | *Worked Example*

Evaluate $\sqrt{-25}$.

Solution.

There is no number, a, such that $a^2 = -25$. So the answer is *This is not a real number.* The square root of a negative number is not a real number. It is a *complex number*, another branch of mathematics that you would study in a future math class.

Note. Careful not to confuse $-\sqrt{4}$ and $\sqrt{-4}$. $-\sqrt{4} = -2$ but $\sqrt{-4}$ is not a real number.

Problem 5 : | *Media/Class Example*

Evaluate the following.

a) $\sqrt{-16}$ b) $-\sqrt{4}$ c) $-\sqrt{-25}$

Problem 6 : | *You Try*

Evaluate the following.

a) $-\sqrt{121}$ b) $\sqrt{-1}$ c) $-\sqrt{81}$

When the radicand is a perfect square, such as 4, 100, or 0.25, the square root evaluates a rational number. What about when the radicand is not a perfect square? The square root of a non-perfect square such as $\sqrt{2}$, is known as an *irrational number*.

The set of **Real** numbers, includes all **Rational** and **Irrational** numbers.

The diagram below summarizes the different sets of real numbers.

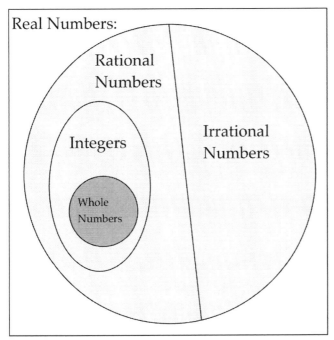

The set of **Rational** numbers, contains:

- Numbers that can be represented as a ratio of integers
- Decimals that either terminate or repeat
  Examples: $\dfrac{1}{2} = 0.5$, $\dfrac{2}{3} = 0.66\bar{6}$, $\dfrac{5}{1} = 5$

The set of **Integers** contain:

- Zero
- *Counting* numbers (1, 2, 3..)
- and their negative values
  Examples: $-3, -1, 0, 1, 3$

The set of **Whole** numbers contains:

- *Counting* numbers
- Zero
  $\{0, 1, 2, 3, 4, ...\}$

The set of **Irrational** numbers:

- Cannot be represented as a ratio of integers
- The decimal representations never terminates nor repeats
- Calculators provide *approximations* only

Ex. $\pi, \sqrt{2}, -\sqrt{5}, e$

Since an irrational number cannot be represented exactly as a decimal, we often leave them in their symbolic form, such as $\sqrt{2}$ or $\sqrt{5}$. If a problem asks for an approximation, then we use our calculator.

Problem 7 : | *Worked Example*

Using a calculator, approximate the following to the nearest hundredth.

a) $\sqrt{2}$ b) $-\sqrt{5}$

   Solution.

   a) Using a calculator that displays 13 decimal places, $\sqrt{2}$ is shown as 1.4142135623731. Rounding to the nearest hundredth gives us $\sqrt{2} \approx 1.41$. Remember to round correctly. We need to look at the thousandths place or the 3rd decimal place to decide whether to round up or down.
   The symbol $\approx$ means *"is approximately."*

   b) Using the same calculator, $-\sqrt{5}$ is shown as -2.23606797749979. Rounding to the nearest hundredth gives us $\sqrt{5} \approx -2.24$.

Problem 8 : | *You Try*

Using a calculator, approximate the following to the nearest hundredth.

a) $\sqrt{6}$ b) $\sqrt{11}$ c) $-\sqrt{5}$

Expressions with Square Roots

Rational and irrational numbers are not alike, so they cannot be combined.

Example. $5 + \sqrt{5}$ is the sum of a rational and an irrational number. These cannot be combined in any way. These are not like terms.

Example. $4\sqrt{7}$ is the product of a rational and an irrational number. This cannot be combined.

However, these expressions can be approximated using a calculator. It is important to save rounding for the last step. It is best to use your calculator for each step; do not write down intermediate values. Remember to round at the very end.

Problem 9 : | *Worked Example*

Approximate $2\sqrt{3}$ to the nearest hundredth.

   Solution.

There is a large assortment of calculators and most of them work differently so be familiar with how your calculator operates. On a calculator that displays 13 decimal places, $2\sqrt{3}$ is shown as 3.46410161513775. Rounding to the nearest hundredth gives us $2\sqrt{3} \approx 3.46$

Problem 10 : | *Media/Class Example*

Approximate $4 + \sqrt{10}$ to the nearest hundredth.

Problem 11 : | *Media/Class Example*

Approximate the following to 3 decimal places.

a) $6\sqrt{2}$

c) $\dfrac{1}{2} - \dfrac{\sqrt{3}}{2}$

b) $5 - \sqrt{30}$

d) $\dfrac{8 - \sqrt{6}}{2}$

Problem 12 : | *You Try*

Approximate the following to 3 decimal places.

a) $3 - \sqrt{17}$

c) $\dfrac{2 - \sqrt{14}}{2}$

b) $10\sqrt{3}$

d) $\sqrt{-21}$

World View Note: The radical sign, when first used was an R with a line through the tail, similar to our perscription symbol today. The R came from the latin, "radix", which can be translated as "source" or "foundation". It wasn't until the 1500s that our current symbol was first used in Germany (but even then it was just a check mark with no bar over the numbers)!

Practice Problems : *Square Roots*

Evaluate the following. If answer is a complex number, state Not Real.

1. $\sqrt{100}$

2. $-\sqrt{64}$

3. $\sqrt{36}$

4. $-\sqrt{121}$

5. $2\sqrt{1}$

6. $\sqrt{-49}$

7. $3+\sqrt{4}$

8. $7-\sqrt{9}$

9. $\sqrt{\dfrac{25}{16}}$

10. $3\sqrt{\dfrac{4}{9}}$

11. $\dfrac{5+\sqrt{16}}{2}$

12. $\dfrac{1-\sqrt{9}}{3}$

Approximate the following to the nearest tenth.

13. $\sqrt{15}$

14. $-\sqrt{-3}$

15. $-\sqrt{21}$

16. $4\sqrt{7}$

17. $-\dfrac{\sqrt{6}}{2}$

18. $4-\sqrt{30}$

19. $10+5\sqrt{3}$

20. $\dfrac{5+\sqrt{3}}{2}$

21. $\dfrac{9-4\sqrt{2}}{5}$

7.2. Symplifying Square Roots

Objective: Simplify expressions containing square roots.

Consider $\sqrt{12}$. This is an irrational number. It cannot be represented as an exact decimal. However, it can be simplified. To do that, we need the following properties.

> **Product Rule for Radicals:** *Given $a > 0$ and $b > 0$, $\sqrt{ab} = \sqrt{a}\,\sqrt{b}$.*

> **Quotient Rule for Radicals:** Given $a > 0$ and $b > 0$, $\sqrt{\dfrac{a}{b}} = \dfrac{\sqrt{a}}{\sqrt{b}}$.

You might recognize that these are very similar to the Product Rule and Quotient Rule for Exponents.

Problem 1 : | Worked Example

Simplify $\sqrt{12}$. Give the exact answer and then approximate to 3 decimal places.

> **Solution.**

The question we want to answer is "Does 12 have a factor that is a perfect square?" If so, we can factor and use the Product Rule for Radicals.

$$\sqrt{12} \qquad \text{Factor 12 such that one factor is a perfect square}$$
$$= \sqrt{4 \cdot 3} \qquad \text{Apply the Product Rule for Radicals}$$
$$= \sqrt{4} \cdot \sqrt{3} \qquad \text{Simplify the perfect square}$$
$$= 2\sqrt{3} \qquad \text{Exact Answer}$$

To approximate, we use the calculator. $2\sqrt{3} \approx 3.464$ rounded to 3 decimal places.

Problem 2 : | Worked Example

Simplify $\sqrt{\dfrac{3}{25}}$. Give the exact answer and then approximate to 2 decimal places.

> **Solution.**

We see that the denominator of the fraction is a perfect square.

$$\sqrt{\frac{3}{25}} \qquad \text{Apply the Quotient Rule for Radicals}$$
$$= \frac{\sqrt{3}}{\sqrt{25}} \qquad \text{Simplify the perfect square}$$
$$= \frac{\sqrt{3}}{5} \qquad \text{Exact Answer}$$

To approximate, we use the calculator. $\dfrac{\sqrt{3}}{5} \approx 0.35$ rounded to 2 decimal places.

Problem 3 : | *Media/Class Example*

Simplify the following. Give the exact answer and then approximate to 2 decimal places.

a) $\sqrt{8}$

c) $\sqrt{\dfrac{50}{9}}$

b) $-\sqrt{72}$

d) $1 + \sqrt{20}$

Problem 4 : | *You Try*

Simplify the following. Give the exact answer and then approximate to 3 decimal places.

a) $\sqrt{\dfrac{2}{49}}$

d) $4 - \sqrt{32}$

b) $-\sqrt{700}$

e) $\sqrt{-24}$

c) $\sqrt{\dfrac{75}{16}}$

f) $8 + \sqrt{18}$

Problem 5 : | *Worked Example*

Simplify $\dfrac{8-\sqrt{12}}{2}$. Give the exact answer and then approximate to 2 decimal places.

> **Solution.**

The first step is to simplify the radical in the numerator. From Problem #1, we see that

$\sqrt{12}=2\sqrt{3}$. Now the problem becomes $\dfrac{8-2\sqrt{3}}{2}$. We can simplify further.

$$\dfrac{8-2\sqrt{3}}{2} \qquad \text{Factor out GCF in the numerator}$$

$$=\dfrac{2\left(4-\sqrt{3}\right)}{2} \qquad \text{Simplify}$$

$$=4-\sqrt{3} \qquad \text{Exact Answer}$$

To approximate, we use the calculator. $4-\sqrt{3}\approx 2.27$ rounded to 2 decimal places.

Note. There is an alternate approach we could use to simplify $\dfrac{8-2\sqrt{3}}{2}$. We can simplify $\dfrac{8-2\sqrt{3}}{2}$ as $\dfrac{8}{2}-\dfrac{2\sqrt{3}}{2}$. (Do you know why?) Then simplify each term.

$$\dfrac{8-2\sqrt{3}}{2} = \dfrac{8}{2}-\dfrac{2\sqrt{3}}{2}$$

$$= 4-\sqrt{3}$$

Problem 6 : | *Media/Class Example*

Simplify $\dfrac{6+\sqrt{8}}{10}$. Give exact answer and then approximate to 1 decimal place.

Problem 7 : | You Try

Simplify the following. Give exact answer and then approximate to 1 decimal place.

a) $\dfrac{3-\sqrt{72}}{3}$
b) $\dfrac{10+\sqrt{20}}{14}$

Squaring a Square Root

What about squaring a square root, such as $\left(\sqrt{7}\right)^2$? Let's see what happens.

$$\left(\sqrt{7}\right)^2 = \left(\sqrt{7}\right)\cdot\left(\sqrt{7}\right)$$
$$= \sqrt{7\cdot 7} \quad \text{(using Product Rule)}$$
$$= \sqrt{49}$$
$$= 7$$

$$\boxed{\left(\sqrt{b}\right)^2 = b \text{ when } b \geqslant 0}$$

Problem 8 : | Media/Class Example

Simplify the following. Give the exact answer and then approximate to 3 decimal places.

a) $2\left(5+\sqrt{3}\right)$
b) $14-\left(\sqrt{20}\right)^2$
c) $6\left(\dfrac{\sqrt{20}}{3}\right)$

| Problem 9 : | *You Try* |

Simplify the following. Give exact answer and then approximate to 3 decimal places.

a) $\left(4\left(\dfrac{5+\sqrt{27}}{4}\right)-5\right)$

b) $\left(2\left(\dfrac{1-\sqrt{3}}{2}\right)-1\right)^{2}$

Practice Problems: *Symplifying Square Roots*

Simplify the following. Give an exact answer, and if irrational, approximate to the nearest hundredth. If answer is a complex number, state Not Real.

1. $\sqrt{8}$

2. $\sqrt{32}$

3. $-\sqrt{45}$

4. $\sqrt{-20}$

5. $\dfrac{\sqrt{72}}{4}$

6. $\sqrt{\dfrac{20}{9}}$

7. $3+\sqrt{50}$

8. $7-\sqrt{12}$

9. $\dfrac{1}{2}+\dfrac{\sqrt{63}}{2}$

10. $\dfrac{6+\sqrt{28}}{2}$

11. $\dfrac{5+\sqrt{24}}{5}$

12. $\dfrac{15-\sqrt{90}}{6}$

CHAPTER 7 ASSESSMENT

Simplify the following expressions.

1. $\sqrt{81}$

2. $3\sqrt{36}$

3. $\sqrt{\dfrac{25}{64}}$

4. $\dfrac{4\sqrt{100}}{25}$

5. $12-\sqrt{16}$

6. $\dfrac{9-\sqrt{49}}{4}$

Approximate the following to the nearest tenth.

7. $\sqrt{22}$

8. $-\sqrt{183}$

9. $\sqrt{\dfrac{2}{5}}$

10. $\dfrac{19+\sqrt{5}}{2}$

Simplify the following. Give exact answers and if irrational, approximate to the nearest hundredth.

11. $\sqrt{12}$

12. $3\sqrt{32}$

13. $\sqrt{\dfrac{11}{25}}$

14. $\dfrac{\sqrt{50}}{5}$

15. $-\sqrt{80}$

16. $\dfrac{6-\sqrt{48}}{4}$

Chapter 8

Graphs of Quadratic Equations

8.1. Square Root Property

Objective: To solve quadratic equations using the square root property.

Square Root Property

In this section, we will see how square roots can be used to find solutions to quadratic equations. The key to this is what is known as the "Square Root Property." To help motivate this property, we first make a few observations; consider the following two equations:

$$\sqrt{(3)^2} = \sqrt{9} = 3$$

$$\sqrt{(-3)^2} = \sqrt{9} = 3$$

More generally, what can we say about $\sqrt{x^2}$? From the above two equations, we observe that:

- when x is a positive real number, $\sqrt{x^2}$ is equal to x itself. From the above equation, $\sqrt{(3)^2} = 3$ since $x = 3$, a positive real number.

- However, when x is a negative real number, $\sqrt{x^2}$ is equal to $-x$. From the above equation, $\sqrt{(-3)^2} = -(-3) = 3$ since $x = -3$.

This implies that the operation $\sqrt{(\ \ )^2}$ has the same effect as taking the absolute value of the number! This is known as the "Square Root Property."

Square Root Property: $\sqrt{x^2} = |x|$

For more general expressions, such as $\sqrt{(x+5)^2}$, we still use the absolute value: $\sqrt{(x+5)^2} = |x+5|$. This is an illustration of the more general square root property:

General Square Root Property: $\sqrt{(A)^2} = |A|$

Problem 1 : | *Worked Example*

Simplify the following expressions.

$$\text{a)} \quad \sqrt{(3)^2}$$

$$\text{c)} \quad \sqrt{z^2}$$

$$\text{b)} \quad \sqrt{(-3)^2}$$

$$\text{d)} \quad \sqrt{(x-7)^2}$$

Solution.

Notice that parts (a) and (b) are the initial two equations from the beginning of the section. We will now simplify these, along with the other expressions, using the Square Root Property.

a) $\sqrt{(3)^2} = |3| = 3$

c) $\sqrt{z^2} = |z|$

b) $\sqrt{(-3)^2} = |-3| = 3$

d) $\sqrt{(x-7)^2} = |x-7|$

Problem 2 : | *Media/Class Example*

Simplify the following expressions.

a) $\sqrt{4^2}$

c) $\sqrt{(5x)^2}$

b) $\sqrt{(-4)^2}$

d) $\sqrt{(2y+3)^2}$

Problem 3 : | *You Try*

Simplify the following expressions.

a) $\sqrt{(-1)^2}$

c) $\sqrt{4x^2}$

b) $\sqrt{(0)^2}$

d) $\sqrt{(7-3y)^2}$

Quadratic Equations

We are now in a position to solve quadratic equations using the Square Root Property. The essential idea is to "undo" the square using a square root; this will then lead to an absolute value equation which we can solve using the techniques from Chapter 1.

| **Problem 4 :** | *Worked Example* |

Solve the following equations and check your solutions. Be sure to simplify your answers. If answers are irrational, round to the nearest hundredth.

a) $x^2 = 4$
b) $x^2 = 7$

Solution.

a)

$$x^2 = 4 \quad \text{Take square root of each side}$$
$$\sqrt{x^2} = \sqrt{4} \quad \text{Apply Square Root Property on left; Simplify on right}$$
$$|x| = 2 \quad \text{Solve absolute value equation}$$
$$x = 2 \text{ or } x = -2 \quad \text{Our Solutions}$$

Check.

When $x = 2$:
When $x = -2$:

$$(2)^2 = 4 \quad \text{Check!} \qquad (-2)^2 = 4 \quad \text{Check!}$$

b)

$$x^2 = 7 \quad \text{Take square root of each side}$$
$$\sqrt{x^2} = \sqrt{7} \quad \text{Apply Square Root Property on left; Simplify on right}$$
$$|x| = 7 \quad \text{Solve absolute value equation}$$
$$x = \sqrt{7} \quad \text{or} \quad x = -\sqrt{7} \quad \text{Exact Solutions}$$
$$x \approx 2.65 \text{ or } x \approx -2.65 \quad \text{Approximate Solutions}$$

Check.

When $x = \sqrt{7}$:
When $x = -\sqrt{7}$:

$$\left(\sqrt{7}\right)^2 = 7 \quad \text{Check!} \qquad \left(-\sqrt{7}\right)^2 = 7 \quad \text{Check!}$$

Note.

1. In equation (a) above, since 4 is a perfect square, we can also solve this equation by factoring:

$$x^2 = 4 \quad \text{Write in standard form}$$
$$x^2 - 4 = 0 \quad \text{Factor left side of equation}$$
$$(x + 2)(x - 2) = 0 \quad \text{Use zero product property}$$
$$x + 2 = 0 \quad \text{or} \quad x - 2 = 0 \quad \text{Solve for } x$$
$$x = 2 \quad \text{or} \quad x = -2 \quad \text{Our solutions}$$

We won't be able to solve equation (b) by factoring, however, since 7 is not a perfect square. Writing the equation $x^2 = 7$ in standard form gives us:

$$x^2 - 7 = 0$$

The polynomial $x^2 - 7$ is prime, and so factoring is not an option here. We need to use square roots!

2. We can condense the form of our solutions to both of these problems using what is known as the "plus and minus" notation: $\pm$. So, for example, instead of writing:

$$x = 2 \quad \text{or} \quad x = -2$$

we can write:

$$x = \pm 2.$$

It is important to remember that when writing something like $x = \pm\sqrt{7}$, you are really specifying two values of x, namely $x = \sqrt{7}$ and $x = -\sqrt{7}$.

3. When checking our answers, use the exact form (i.e., the simplified radical form) rather than the approximate form (i.e., the number that was rounded from the calculator output) to avoid discrepancies. In example (b), if we had checked using $x = 2.65$ rather than $x = \sqrt{7}$, this is what we would have gotten:

$$(2.65)^2 = 7.0025$$
$$\neq 7$$

Since $\sqrt{7}$ was approximated to 2.65, we will not get exactly 7 when 2.65 is squared although the answer, 7.0025 is close to 7. To get exactly 7, we will need to use $x = \sqrt{7}$ to check our answer.

More generally, we have the following useful property we will use from now on:
If $|A| = B$, then $A = \pm B$.

Problem 5 : | *Worked Example*

Solve $x^2 = 18$ and check your solution. Be sure to simplify your answers. If answers are irrational, round to the nearest hundredth.

Solution.

$$
\begin{aligned}
x^2 &= 18 \quad &&\text{Take square root of each side} \\
\sqrt{x^2} &= \sqrt{18} \quad &&\text{Apply Square Root Property on left; Simplify on right} \\
|x| &= 3\sqrt{2} \quad &&\text{Solve absolute value equation} \\
x &= \pm 3\sqrt{2} \quad &&\text{Exact Solutions} \\
x &\approx \pm 4.24 \quad &&\text{Approximate Solutions}
\end{aligned}
$$

Check when $x = 3\sqrt{2}$:

$$\left(3\sqrt{2}\right)^2 \stackrel{?}{=} 18$$
$$9 \cdot 2 = 18 \qquad \text{Check!}$$

Check when $x = -3\sqrt{2}$:

$$\left(-3\sqrt{2}\right)^2 \stackrel{?}{=} 18$$
$$9 \cdot 2 = 18 \qquad \text{Check!}$$

Problem 6 : | *Worked Example*

Solve $(x-3)^2 = 11$ and check your solution. Be sure to simplify your answers. If answers are irrational, round to the nearest tenth.

Solution.

$$(x-3)^2 = 11 \quad \text{Take square root of each side}$$
$$\sqrt{(x-3)^2} = \sqrt{11} \quad \text{Apply Square Root Property on left; Simplify on right}$$
$$|x-3| = \sqrt{11} \quad \text{Solve absolute value equation}$$
$$x - 3 = \pm\sqrt{11} \quad \text{Add 3 to each side of equation}$$
$$x = 3 \pm \sqrt{11} \quad \text{Exact Solutions}$$
$$x \approx 6.3 \text{ or } x \approx -0.3 \quad \text{Approimate Solutions}$$

Check when $x = 3 + \sqrt{11}$:

$$\left[\left(3+\sqrt{11}\right)-3\right]^2 \stackrel{?}{=} 11$$
$$\left[3+\sqrt{11}-3\right]^2 \stackrel{?}{=} 11$$
$$\left[\sqrt{11}\right]^2 = 11 \qquad \text{Check!}$$

Check when $x = 3 - \sqrt{11}$:

$$\left[\left(3-\sqrt{11}\right)-3\right]^2 \stackrel{?}{=} 11$$
$$\left[3-\sqrt{11}-3\right]^2 \stackrel{?}{=} 11$$
$$\left[-\sqrt{11}\right]^2 = 11 \qquad \text{Check!}$$

Problem 7 : | *Worked Example*

Solve $(x+4)^2 = 9$ and check your solution. Be sure to simplify your answers. If answers are irrational, round to the nearest thousandth.

Solution.

$$
\begin{array}{ll}
(x+4)^2 = 9 & \text{Take square root of each side} \\
\sqrt{(x+4)^2} = \sqrt{9} & \text{Apply Square Root Property on left; Simplify on right} \\
|x+4| = 3 & \text{Solve absolute value equation} \\
x+4 = \pm 3 & \text{Subtract 4 from each side of equation} \\
x = -4 \pm 3 & \text{Rational answers; Perform indicated operation} \\
x = -4+3 \text{ or } x = -4-3 & \text{Simplify} \\
x = -1 \quad \text{or} \quad x = -7 & \text{Our Solutions}
\end{array}
$$

Check when $x = -1$:

$$
\begin{array}{l}
(-1+4)^2 \stackrel{?}{=} 9 \\
\qquad (3)^2 = 9 \qquad \text{Check!}
\end{array}
$$

Check when $x = -7$:

$$
\begin{array}{l}
(-7+4)^2 \stackrel{?}{=} 9 \\
\qquad (-3)^2 = 9 \qquad \text{Check!}
\end{array}
$$

Problem 8 : | *Worked Example*

Solve $(2x-1)^2 = 7$ and check your solution. Be sure to simplify your answers. If answers are irrational, round to the nearest hundredth.

Solution.

$$
\begin{array}{ll}
(2x-1)^2 = 7 & \text{Take square root of each side} \\
\sqrt{(2x-1)^2} = \sqrt{7} & \text{Apply Square Root Property on left; Simplify on right} \\
|2x-1| = \sqrt{7} & \text{Solve absolute value equation} \\
2x-1 = \pm\sqrt{7} & \text{Add 1 to each side of equation} \\
2x = 1 \pm \sqrt{7} & \text{Divide each side by 2} \\
x = \dfrac{1 \pm \sqrt{7}}{2} & \text{Exact Solutions} \\
x \approx 1.82 \text{ or } x \approx -0.82 & \text{Approximate Solutions}
\end{array}
$$

Check when $x = \dfrac{1 + \sqrt{7}}{2}$:

$$\left[2\left(\dfrac{1+\sqrt{7}}{2} \right) - 1 \right]^2 \overset{?}{=} 7$$

$$\left[\left(1+\sqrt{7} \right) - 1 \right]^2 \overset{?}{=} 7$$

$$\left[1 + \sqrt{7} - 1 \right]^2 \overset{?}{=} 7$$

$$\left[\sqrt{7} \right]^2 = 7 \qquad \text{Check!}$$

Check when $x = \dfrac{1 - \sqrt{7}}{2}$:

$$\left[2\left(\dfrac{1-\sqrt{7}}{2} \right) - 1 \right]^2 \overset{?}{=} 7$$

$$\left[\left(1-\sqrt{7} \right) - 1 \right]^2 \overset{?}{=} 7$$

$$\left[1 - \sqrt{7} - 1 \right]^2 \overset{?}{=} 7$$

$$\left[-\sqrt{7} \right]^2 = 7 \qquad \text{Check!}$$

Problem 9 : | *Worked Example*

Solve $(x+6)^2 = -4$ and check your solution. Be sure to simplify your answers. If answers are irrational, round to the nearest hundredth.

Solution.

$$(x+6)^2 = -4 \quad \text{Take square root of each side}$$

$$\sqrt{(x+6)^2} = \sqrt{-4} \quad \sqrt{-4} \text{ is a complex number!}$$

Thus, this equation does not possess any real solutions. Solutions are complex numbers.

We will learn about these numbers in the next math class.

Problem 10 : | *Media/Class Example*

Solve the following equations and check your solutions. Be sure to simplify your answers. If answers are irrational, round to the nearest thousandth.

a) $m^2 = \dfrac{2}{9}$

c) $(3x+2)^2 = 36$

b) $(y-5)^2 = 24$

d) $(g+8)^2 = -16$

Problem 11 : | *You Try*

Solve the following equations and check your solutions. Be sure to simplify your answers. If answers are irrational, round to the nearest hundredth.

a) $x^2 = \dfrac{3}{4}$

c) $(p-2)^2 = 8$

b) $(y+7)^2 = 49$

d) $(2w+4)^2 = 12$

Here are some examples that require us to first isolate the square term and then apply the Square Root Property.

| Problem 12 : | Worked Example |

Solve $2x^2 - 1 = 17$ and check your solutions. Be sure to simplify your answers. If answers are irrational, round to the nearest thousandth.

Solution.

$$\begin{aligned}
2x^2 - 1 &= 17 &&\text{Isolate } x^2 \text{ term by adding 1 to each side}\\
2x^2 &= 18 &&\text{Divide each side by 2}\\
x^2 &= 9 &&\text{Take square root of each side}\\
\sqrt{x^2} &= \sqrt{9} &&\text{Apply square root property on left; Simplify on right}\\
|x| &= 3 &&\text{Solve absolute value equation}\\
x &= \pm 3 &&\text{Our solutions}
\end{aligned}$$

Check when $x = 3$:

$$\begin{aligned}
2(3)^2 - 1 &\overset{?}{=} 17\\
2(9) - 1 &\overset{?}{=} 17\\
18 - 1 &= 17 &&\text{Check!}
\end{aligned}$$

Check when $x = -3$:

$$\begin{aligned}
2(-3)^2 - 1 &\overset{?}{=} 17\\
2(9) - 1 &\overset{?}{=} 17\\
18 - 1 &= 17 &&\text{Check!}
\end{aligned}$$

| Problem 13 : | Worked Example |

Solve $(x - 4)^2 - 3 = 5$ and check your solutions. Be sure to simplify your answers. If answers are irrational, round to the nearest hundredth.

Solution.

$$\begin{aligned}
(x-4)^2 - 3 &= 5 &&\text{Isolate } (x-4)^2 \text{ term by adding 3 to each side}\\
(x-4)^2 &= 8 &&\text{Take square root of each side}\\
\sqrt{(x-4)^2} &= \sqrt{8} &&\text{Apply square root property on left; Simplify on right}\\
|x-4| &= 2\sqrt{2} &&\text{Solve absolute value equation}\\
x-4 &= \pm 2\sqrt{2} &&\text{Add 4 to each side of equation}\\
x &= 4 \pm 2\sqrt{2} &&\text{Exact Solutions}\\
x &\approx 6.83 \text{ or } x \approx 1.17 &&\text{Approximate Solutions}
\end{aligned}$$

Check when $x = 4 + 2\sqrt{2}$:

$$\left[\left(4 + 2\sqrt{2}\right) - 4\right]^2 - 3 \stackrel{?}{=} 5$$

$$\left[4 + 2\sqrt{2} - 4\right]^2 - 3 \stackrel{?}{=} 5$$

$$\left[2\sqrt{2}\right]^2 - 3 \stackrel{?}{=} 5$$

$$4(2) - 3 \stackrel{?}{=} 5$$

$$8 - 3 = 5 \qquad \text{Check!}$$

Check when $x = 4 - 2\sqrt{2}$:

$$\left[\left(4 - 2\sqrt{2}\right) - 4\right]^2 - 3 \stackrel{?}{=} 5$$

$$\left[4 - 2\sqrt{2} - 4\right]^2 - 3 \stackrel{?}{=} 5$$

$$\left[-2\sqrt{2}\right]^2 - 3 \stackrel{?}{=} 5$$

$$4(2) - 3 \stackrel{?}{=} 5$$

$$8 - 3 = 5 \qquad \text{Check!}$$

Problem 14 : | *Worked Example*

Solve $(x - 4)^2 - 3 = -5$ and check your solutions. Be sure to simplify answers. If answers are irrational, round to the nearest hundredth.

Solution.

$$(x - 4)^2 - 3 = -5 \quad \text{Isolate } (x - 4)^2 \text{ term by adding 3 to each side}$$

$$(x - 4)^2 = -2 \quad \text{Take square root of each side}$$

$$\sqrt{(x - 4)^2} = \sqrt{-2} \quad \sqrt{-2} \text{ is a complex number}$$

Equation does not have any real solutions.

Problem 15 : | *Worked Example*

Solve $3 - 4(6x + 5)^2 = -33$ and check your solutions. Be sure to simplify your answers. If answers are irrational, round to the nearest tenth.

Solution.

$$3 - 4(6x + 5)^2 = -33 \quad \text{Isolate } (6x + 5)^2 \text{ term by adding } -3 \text{ to each side}$$

$$-4(6x + 5)^2 = -36 \quad \text{Divide each side by } -4$$

$$(6x + 5)^2 = 9 \quad \text{Take square root of each side}$$

$$\sqrt{(6x + 5)^2} = \sqrt{9} \quad \text{Apply square root property on left; Simplify on right}$$

$$|6x + 5| = 3 \quad \text{Solve absolute value equation}$$

$$6x + 5 = \pm 3 \quad \text{Add } -5 \text{ to each side of equation}$$
$$6x = -5 \pm 3 \quad \text{Divide each side by 6}$$

$$x = \frac{-5 \pm 3}{6} \quad \text{Rational answers; Perform indicated operation}$$

$$x = \frac{-5 + 3}{6} \text{ or } x = \frac{-5 - 3}{6} \quad \text{Perform indicated operation and simplify}$$

$$x = \frac{-2}{6} \quad \text{or} \quad x = \frac{-8}{6} \quad \text{Simplify}$$

$$x = -\frac{1}{3} \quad \text{or} \quad x = -\frac{4}{3} \quad \text{Our Solutions}$$

Check when $x = -\frac{1}{3}$:

$$3 - 4\left[6\left(-\frac{1}{3}\right) + 5\right]^2 \overset{?}{=} -33$$
$$3 - 4[-2 + 5]^2 \overset{?}{=} -33$$
$$3 - 4[3]^2 \overset{?}{=} -33$$
$$3 - 4(9) \overset{?}{=} -33$$
$$3 - 36 = -33 \quad \text{Check!}$$

Check when $x = -\frac{4}{3}$:

$$3 - 4\left[6\left(-\frac{4}{3}\right) + 5\right]^2 \overset{?}{=} -33$$
$$3 - 4[-8 + 5]^2 \overset{?}{=} -33$$
$$3 - 4[-3]^2 \overset{?}{=} -33$$
$$3 - 4(9) \overset{?}{=} -33$$
$$3 - 36 = -33 \quad \text{Check!}$$

Problem 16 : | *Media/Class Example*

Solve the following euations and check your solutions. Be sure to simplify your answers. If answers are irrational, round to the nearest tenth.

a) $(x-6)^2+3=5$ c) $3-5p^2=8$

b) $4y^2-1=6$ d) $5-3(2c-1)^2=-7$

Problem 17 : | *You Try*

Solve the following euations and check your solutions. Be sure to simplify your answers. If answers are irrational, round to the nearest thousandth.

a) $4-3x^2=1$ c) $3-5p^2=7p^2+9$

b) $2(y+3)^2+4=5$ d) $3(6z-8)^2-5=91$

Practice Problems: *Square Root Property*

Fill in the blanks.

1. $\sqrt{(\square)^2} =$ _____ is known as the _____ Property.

Simplify the following expressions.

2. $\sqrt{(7)^2}$

3. $\sqrt{(-7)^2}$

4. $\sqrt{A^2}$

5. $\sqrt{(A+B)^2}$

6. $\sqrt{(x+3)^2}$

7. $\sqrt{(2y-9)^2}$

8. $\sqrt{x^2+6x+9}$

Solve each equation. Be sure to fully simplify and check each of your solutions using the exact form of your answer. If answers are irrational, round to the nearest hundredth.

9. $x^2 = 36$

10. $y^2 = 11$

11. $z^2 = 32$

12. $w^2 = -4$

13. $x^2 = 13$

14. $x^2 + 4 = 9$

15. $3y^2 = 18$

16. $2y^2 = 144$

17. $3z^2 - 29 = -5$

18. $\dfrac{1}{10}v^2 = 30$

19. $5 - 2t^2 = 7t^2 + 4$

20. $(x-3)^2 = 7$

21. $(z-1)^2 = 2$

22. $(y+4)^2 + 3 = 52$

23. $4 - (h+2)^2 = 0$

24. $(2t+5)^2 = 10$

25. $(4y+8)^2 = 12$

26. $(3g-6)^2 - 18 = 0$

27. $8 - (2p+1)^2 = -1$

28. $100t^2 - 400 = 2000$

29. $5 - 3x^2 = 6 - 2x^2$

30. $5(x-8)^2 = 25$

31. $2(3y+4)^2 - 7 = 9$

8.2. COMPLETING THE SQUARE

Objective: Solve quadratic equations by completing the square.

We will learn another method of solving quadratic equations known as completing the square. The idea is to change the quadratic expression into a perfect square trinomial and then use the square root property to solve for the variable. Let us first look at an example at how this works.

| **Problem 1 :** | *Worked Example* |

Solve $x^2 + 2x + 1 = 7$. Leave the answer in exact and simplified form.

> **Solution.**

The idea behind solving a quadratic by completing the square is to have a perfect square trinomial. Recall from the previous chapter how we can recognize a perfect square trinomial. Notice that the quadratic term, x^2, and the constant term, 1, are perfect squares. The square root of the quadratic term is x and the square root of the constant term is 1. When multiplied together, the product is $1x$ or x. When x is multiplied by 2, we get the linear term, $2x$, which is also the linear term of the trinomial. This indicates that the trinomial, $x^2 + 2x + 1$, is a perfect square trinomial and factors as $(x+1)^2$. We can now use the method from the previous section to solve the quadratic equation.

$$
\begin{aligned}
x^2 + 2x + 1 &= 7 \quad && \text{Factor perfect square trinomial} \\
(x+1)^2 &= 7 \quad && \text{Take square root of each side} \\
\sqrt{(x+1)^2} &= \sqrt{7} \quad && \text{Apply Square Root Property on left; Simplify on right} \\
|x+1| &= \sqrt{7} \quad && \text{Solve absolute value equation} \\
x+1 &= \pm\sqrt{7} \quad && \text{Subtract 1 on each side of equation} \\
x &= -1 \pm \sqrt{7} \quad && \text{Our Solutions}
\end{aligned}
$$

The trinomial in the above example is already a perfect square trinomial. Unfortunately, that situation does not occur often. Our goal then is to get the trinomial to become a perfect square trinomial.

We will first practice changing quadratic expressions of the form $x^2 + bx$ into a perfect square trinomial. To do so, we will be searching for the constant term, c, to add to the quadratic expression in order to make it a perfect square trinomial. This is done by squaring half of the linear term, that is $c = \left(\dfrac{1}{2} \cdot b\right)^2$.

Let us see where that formula comes from. Suppose we have:

$$(x+p)^2 = x^2 + 2px + p^2$$

The trinomial is similar in form to the general quadratic expression: $x^2 + bx + c$, where $2p = b$ and $p^2 = c$. Solving for p, we get:

$$2p = b \quad \text{Divide each side by 2}$$

$$p = \frac{1}{2} \cdot b \quad \text{Square each side}$$

$$p^2 = \left(\frac{1}{2} \cdot b\right)^2 \quad \text{Substitute into } c = p^2$$

$$c = \left(\frac{1}{2} \cdot b\right)^2 \quad \text{Our formula}$$

Another way to look at how the constant term, c, is derived is from a geometric standpoint. Our goal is to change the expression $x^2 + bx$ into a perfect square trinomial by completing the square.

x^2 means that we have a figure whose length and width are both x. Only a square can have equal length and width. bx means that we have a figure whose length is b and width is x. This must be a rectangle.

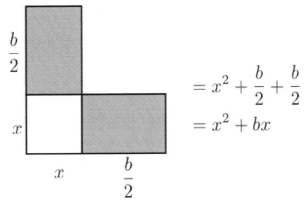

Next, cut side b in half so we can connect the ends as follows.

$$= x^2 + \frac{b}{2} + \frac{b}{2}$$

$$= x^2 + bx$$

We now need to "complete the square" of the above figure by finding the area of the missing portion in question.

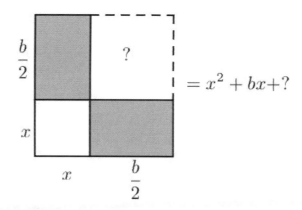

The missing portion in question has sides of length, $\left(\frac{1}{2} \cdot b\right)$ which gives us an area $= \left(\frac{1}{2} \cdot b\right)\left(\frac{1}{2} \cdot b\right) = \left(\frac{1}{2} \cdot b\right)^2$. This is the contant term, c, that needs to be added to the expression $x^2 + bx$ in order to form a perfect square trinomial and also to "complete the square" of the figure. Therefore, $c = \left(\frac{1}{2} \cdot b\right)^2$.

Problem 2 : | *Worked Example*

Find the constant term, c, to make $x^2 + 8x + c$ into a perfect square trinomial. Then factor the trinomial.

> **Solution.**

$$x^2 + 8x + c \quad \text{Find } c; c = \left(\frac{1}{2} \cdot b\right)^2 = \left(\frac{1}{2} \cdot 8\right)^2 = (4)^2 = 16$$
$$x^2 + 8x + 16 \quad \text{Factor perfect square trinomial}$$
$$(x + 4)^2 \quad \text{Our Solution}$$

Check to verify factor is correct.

$$
\begin{aligned}
(x + 4)^2 &= (x + 4)(x + 4) \\
&= x(x + 4) + 4(x + 4) \\
&= x^2 + 4x + 4x + 16 \\
&= x^2 + 8x + 16 \qquad \text{Confirmed!}
\end{aligned}
$$

Let us see how this problem can be done geometrically.

$$\square + \blacksquare = x^2 + 8x$$

Cut the side with length 8 units in half and connect the ends as follows.

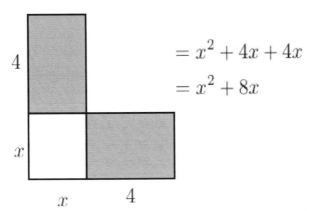

We now need to "complete the square" of the above figure by finding the area of the missing portion in question.

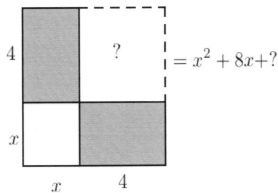

The missing portion in question has sides of length 4. Its area is $4 \cdot 4 = 16$. Therefore 16 must to be added to $x^2 + 8x$ in order to form a perfect square trinomial and also to "complete the square" of the figure.

Problem 3 : | *Worked Example*

Find the constant term, c, to make $p^2 - 16p + c$ into a perfect square trinomial. Then factor the trinomial.

Solution.

$$p^2 - 16p + c \quad \text{Find } c; c = \left(\frac{1}{2} \cdot b\right)^2 = \left(\frac{1}{2} \cdot -16\right)^2 = (-8)^2 = 64$$

$$p^2 - 16p + 64 \quad \text{Factor perfect square trinomial}$$

$$(x - 8)^2 \quad \text{Our Solution}$$

Check to verify factor is correct.

$$\begin{aligned}
(x+4)^2 &= (x+4)(x+4) \\
&= x(x+4) + 4(x+4) \\
&= x^2 + 4x + 4x + 16 \\
&= x^2 + 8x + 16 \quad \text{Confirmed!}
\end{aligned}$$

Problem 4 : | *Worked Example*

Find the constant term, c, to make $y^2 + 7y + c$ into a perfect square trinomial. Then factor the trinomial.

Solution.

$$y^2 + 7y + c \quad \text{Find } c; c = \left(\frac{1}{2} \cdot b\right)^2 = \left(\frac{1}{2} \cdot 7\right)^2 = \left(\frac{7}{2}\right)^2 = \frac{49}{4}$$

$$y^2 + 7y + \frac{49}{4} \quad \text{Factor perfect square trinomial}$$

$$\left(y + \frac{7}{2}\right)^2 \quad \text{Our Solution}$$

Check to verify factor is correct.

$$\left(y + \frac{7}{2}\right)^2 = \left(y + \frac{7}{2}\right)\left(y + \frac{7}{2}\right)$$

$$= y\left(y + \frac{7}{2}\right) + \frac{7}{2}\left(y + \frac{7}{2}\right)$$

$$= y^2 + \frac{7}{2}y + \frac{7}{2}y + \frac{49}{4}$$

$$= y^2 + 7y + \frac{49}{4} \qquad \text{Confirmed!}$$

Problem 5 : | *Worked Example*

Find the constant term, c, to make $z^2 - \dfrac{5}{3}z + c$ into a perfect square trinomial. Then factor the trinomial.

Solution.

$$z^2 - \frac{5}{3}z + c \quad \text{Find } c; c = \left(\frac{1}{2} \cdot b\right)^2 = \left(\frac{1}{2} \cdot -\frac{5}{3}\right)^2 = \left(-\frac{5}{6}\right)^2 = \frac{25}{36}$$

$$z^2 - \frac{5}{3}z + \frac{25}{36} \quad \text{Factor perfect square trinomial}$$

$$\left(z - \frac{5}{6}\right)^2 \quad \text{Our Solution}$$

Check to verify factor is correct.

$$\left(z - \frac{5}{6}\right)^2 = \left(z - \frac{5}{6}\right)\left(z - \frac{5}{6}\right)$$

$$= z\left(x - \frac{5}{6}\right) - \frac{5}{6}\left(z - \frac{5}{6}\right)$$

$$= z^2 - \frac{5}{6}z - \frac{5}{6}z + \frac{25}{36}$$

$$= z^2 - \frac{5}{3}z + \frac{25}{36} \qquad \text{Confirmed!}$$

Problem 6 : | *Media/Class Example*

Find the constant term, c, to make each expression into a perfect square trinomial. Then factor the trinomial.

a) $x^2 + 2x + c$

b) $y^2 - 6y + c$

c) $m^2 - 5m + c$

d) $n^2 + \dfrac{2}{3}n + c$

Problem 7 : | *You Try*

Find the constant term, c, to make each expression into a perfect square trinomial. Then factor the trinomial.

a) $x^2 + 12x + c$

b) $y^2 - 8y + c$

c) $m^2 + 3m + c$

d) $n^2 - \dfrac{3}{5}n + c$

Steps in Solving a Quadratic Equation of the form, $ax^2 + bx + c = 0$ by Completing the Square:

1. Separate constant term from variable terms. Be sure variable terms are written in descending order.

2. If $a \neq 1$, divide each term by a. Otherwise, go to next step.

3. Find the value to complete the square by doing: $\left(\dfrac{1}{2} \cdot b\right)^2$. Add the value to each side of the equation.

4. Factor the trinomial, on one side and add the constant terms, on the other side. Trinomial should factor as a perfect square binomial.

5. Solve for the variable by taking the square root of each side and using the square root property.

Problem 8 : | *Worked Example*

Solve $x^2 + 6x = 16$ by completing the square. Be sure to simplify your answers. If answers are irrational, round to the nearest tenth.

Solution.

Let us go through each of the above steps using this example.

1. Constant and variable terms already separated $\qquad x^2 + 6x \;=\; 16$

2. $a = 1$; move to next step

3. Find $\left(\dfrac{1}{2} \cdot b\right)^2 = \left(\dfrac{1}{2} \cdot 6\right)^2 = (3)^2 = 9$

    Add 9 to each side of equation $\qquad\qquad\qquad x^2 + 6x + 9 \;=\; 16 + 9$

4. Factor trinomial; Add constant terms $\qquad\qquad (x+3)^2 \;=\; 25$

5. Take square root of each side $\qquad\qquad \sqrt{(x+3)^2} \;=\; \sqrt{25}$

    Solve absolute value equation $\qquad\qquad |x+3| \;=\; 5$

    Solve for x $\qquad\qquad\qquad\qquad\qquad x+3 \;=\; \pm 5$

    Rational answers; Perform indicated operation $\qquad x \;=\; -3 \pm 5$

    $\qquad\qquad\qquad\qquad\qquad\qquad x = -3 + 5 \text{ or } x = -3 - 5$

    Our Solutions $\qquad\qquad\qquad\qquad x = 2 \text{ or } x = -8$

Check when $x = 2$:

$$(2)^2 + 6(2) - 16 \overset{?}{=} 0$$
$$4 + 12 - 16 \overset{?}{=} 0$$
$$16 - 16 = 0 \qquad \text{Check!}$$

Check when $x = -8$:

$$(-8)^2 + 6(-8) - 16 \overset{?}{=} 0$$
$$64 - 48 - 16 \overset{?}{=} 0$$
$$16 - 16 = 0 \qquad \text{Check!}$$

Note. The above example can also be solved by factoring.

$$x^2 + 6x - 16 = 0 \quad \text{Factor left-hand side}$$
$$(x + 8)(x - 2) = 0 \quad \text{Use zero product property}$$
$$x + 8 = 0 \quad \text{or} \quad x - 2 = 0 \quad \text{Solve for } x$$
$$x = -8 \quad \text{or} \quad x = 2 \quad \text{Our solutions}$$

From the above example, we see that solving the quadratic equation by factoring or completing the square yields the same answer but factoring is much simpler. However, not every quadratic equation can be solved by factoring. Another tool needs to be used. In this section, we will solve quadratic equations by completing the square.

World View Note: The Babylonians (around 400 BC) developed an algorithmic approach to solving quadratic equations by completing the square. However, only positive solutions were considered. Their solution is geometric in nature. Let us see how they solved $x^2 + 6x = 16$.

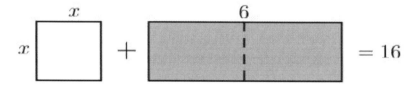

Cut the side with length 6 in half and connect the ends as follows.

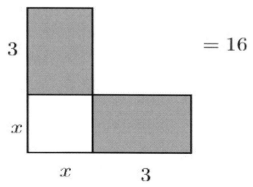

We now need to "complete the square" of the above figure by finding the area of the missing portion in question.

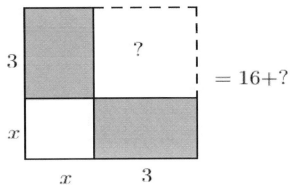

The missing portion in question has sides of length 3. Its area is $3 \cdot 3 = 9$. Therefore 9 must be added on the left to "complete the square." However, since we are working with an equation, in order to balance the equation, if 9 is added on the left, 9 must also be added on the right.

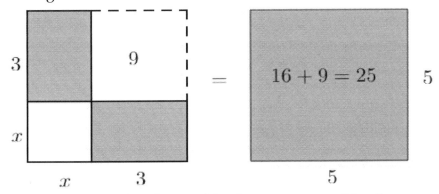

Both figures are equal to each other and both are squares. The figure on the left is a square with sides $(x+3)$ and the figure on the right is a square with length 5. Equating the two sides, we get:

$$x+3 = 5$$
$$x = 2$$

The positive solution to the quadratic equation $x^2 + 6x = 16$ is $x = 2$.

Solving quadratic equations this way focuses on the dimensions of a square. Hence the term "quadratic" which means square.

Problem 9 : | Worked Example

Solve $x^2 - 8x - 2 = 0$ by completing the square. Be sure to simplify your answers. If answers are irrational, round to the nearest hundredth.

Solution.

Note that the trinomial cannot be factored. We will complete the square in order to solve the trinomial. Let us go through the steps.

1. Separate constant and variable terms

      by adding 2 to each side of equation $\qquad$ $x^2 - 8x = 2$

2. $a = 1$; move to next step

3. Find $\left(\frac{1}{2} \cdot b\right)^2 = \left(\frac{1}{2} \cdot -8\right)^2 = (-4)^2 = 16$

      Add 16 to each side of equation $\qquad$ $x^2 - 8x + 16 = 2 + 16$

4. Factor trinomial; Add constant terms $\qquad$ $(x - 4)^2 = 18$

5. Take square root of each side $\qquad$ $\sqrt{(x-4)^2} = \sqrt{18}$

    Solve absolute value equation $\qquad$ $|x - 4| = 3\sqrt{2}$

    Solve for x $\qquad$ $x - 4 = \pm 3\sqrt{2}$

         Exact Solutions $\qquad$ $x = 4 \pm 3\sqrt{2}$

      Approximate Solutions $\qquad$ $x \approx 8.24 \text{ or } -0.24$

| Problem 10 : | *Worked Example* |

Solve $3y^2 + 6y - 12 = 0$ by completing the square. Be sure to simplify your answers. If answers are irrational, round to the nearest thousandth.

   Solution.

Let us go through the steps.

1. Separate constant and variable terms
   by adding 12 to each side of equation $3y^2 + 6y = 12$

2. $a \neq 1$; Divide each term by $a = 3$ $y^2 - 2y = 4$

3. Find $\left(\dfrac{1}{2} \cdot b\right)^2 = \left(\dfrac{1}{2} \cdot -2\right)^2 = (-1)^2 = 1$

   Add 1 to each side of equation $y^2 - 2y + 1 = 4 + 1$

4. Factor trinomial; Add constant terms $(y-1)^2 = 5$

5. Take square root of each side $\sqrt{(y-1)^2} = \sqrt{5}$

   Solve absolute value equation $|y-1| = \sqrt{5}$

   Solve for y $y - 1 = \pm\sqrt{5}$

   Exact Solutions $y = 1 \pm \sqrt{5}$

   Approximate Solutions $y \approx 3.236 \text{ or } -1.236$

| Problem 11 : | *Media/Class Example* |

Solve each quadratic equation by completing the square. Be sure to simplify your answers. If answers are irrational, round to the nearest thousandth.

a) $x^2 + 4x - 5 = 0$ c) $2g^2 + 16g = 4$

b) $y^2 - 12y + 9 = 2$ d) $5n^2 = 10n + 15$

Problem 12 : | *You Try*

Solve each quadratic equation by completing the square. Be sure to simplify your answers. If answers are irrational, round to the nearest hundredth.

a) $x^2 - 6x - 7 = 0$

c) $6p^2 - 48p + 42 = 0$

b) $y^2 + 2y = 1$

d) $2n^2 + 4n - 28 = -6$

Solving quadratic equations by completing the square will sometimes involve fractions. You will need to be comfortable with operations involving fractions.

Problem 13 : | *Worked Example*

Solve $x^2 - 3x - 2 = 0$ by completing the square. Be sure to simplify your answers. If answers are irrational, round to the nearest hundredth.

Solution.

Let us go through the steps.

1. Separate constant and variable terms

      by adding 2 to each side of equation $x^2 - 3x = 2$

2. $a = 1$; move to next step

3. Find $\left(\dfrac{1}{2} \cdot b\right)^2 = \left(\dfrac{1}{2} \cdot -3\right)^2 = \left(-\dfrac{3}{2}\right)^2 = \dfrac{9}{4}$

    Add $\dfrac{9}{4}$ to each side of equation $x^2 - 3x + \dfrac{9}{4} = 2 + \dfrac{9}{4}$

4. Factor trinomial; Add constant terms $\left(x - \dfrac{3}{2}\right)^2 = \dfrac{17}{4}$

5. Take square root of each side $\sqrt{\left(x - \dfrac{3}{2}\right)^2} = \sqrt{\dfrac{17}{4}}$

    Solve absolute value equation $\left|x - \dfrac{3}{2}\right| = \dfrac{\sqrt{17}}{2}$

    Solve for x $x - \dfrac{3}{2} = \pm\dfrac{\sqrt{17}}{2}$

    Exact Solutions $x = \dfrac{3}{2} \pm \dfrac{\sqrt{17}}{2} = \dfrac{3 \pm \sqrt{17}}{2}$

    Approximate Solutions $x \approx 3.56$ or -0.56

Problem 14 : | *Worked Example*

Solve $3y^2 = 5 - 2y$ by completing the square. Be sure to simplify your answers. If answers are irrational, round to the nearest tenth.

Solution.

Let us go through the steps.

1. Separate constant and variable terms
   by adding $2y$ to each side of equation $\qquad\qquad 3y^2 + 2y = 5$

2. $a \neq 1$; Divide each term by $a = 3$ $\qquad\qquad y^2 + \dfrac{2}{3}y = \dfrac{5}{3}$

3. Find $\left(\dfrac{1}{2} \cdot b\right)^2 = \left(\dfrac{1}{2} \cdot \dfrac{2}{3}\right)^2 = \left(\dfrac{1}{3}\right)^2 = \dfrac{1}{9}$

   Add 1 to each side of equation $\qquad\qquad y^2 + \dfrac{2}{3}y + \dfrac{1}{9} = \dfrac{5}{3} + \dfrac{1}{9}$

4. Factor trinomial; Add constant terms $\qquad\qquad \left(y + \dfrac{1}{3}\right)^2 = \dfrac{16}{9}$

5. Take square root of each side $\qquad\qquad \sqrt{\left(y + \dfrac{1}{3}\right)^2} = \sqrt{\dfrac{16}{9}}$

   Solve absolute value equation $\qquad\qquad \left|y + \dfrac{1}{3}\right| = \dfrac{4}{3}$

   Solve for y $\qquad\qquad y + \dfrac{1}{3} = \pm\dfrac{4}{3}$

   Rational answers; Perform indicated operation $\qquad\qquad y = -\dfrac{1}{3} \pm \dfrac{4}{3}$

$$y = -\dfrac{1}{3} + \dfrac{4}{3} \text{ or } y = -\dfrac{1}{3} - \dfrac{4}{3}$$

   Our Solutions $\qquad\qquad y = -1 \text{ or } y = \dfrac{5}{3}$

Problem 15 : | *Worked Example*

Solve $2x^2 = 3x - 7$ by completing the square. Be sure to simplify your answers. If answers are irrational, also round to the nearest thousandth.

Solution.

Let us go through the steps.

1. Separate constant and variable terms

   by subtracting $3x$ from side of equation $\qquad$ $2x^2 - 3x = -7$

2. $a \pm 1$; Divide each term by $a = 2$ $\qquad$ $x^2 - \dfrac{3}{2}x = -\dfrac{7}{2}$

3. Find $\left(\dfrac{1}{2} \cdot b\right)^2 = \left(\dfrac{1}{2} \cdot -\dfrac{3}{2}\right)^2 = \left(-\dfrac{3}{4}\right)^2 = \dfrac{9}{16}$

   Add $\dfrac{9}{16}$ to each side of equation $\qquad$ $x^2 - \dfrac{3}{2}x + \dfrac{9}{16} = -\dfrac{7}{2} + \dfrac{9}{16}$

4. Factor trinomial; Add constant terms $\qquad$ $\left(x - \dfrac{3}{4}\right)^2 = -\dfrac{47}{4}$

5. Take square root of each side $\qquad$ $\sqrt{\left(x - \dfrac{3}{4}\right)^2} = \sqrt{-\dfrac{47}{4}}$

$\sqrt{-\dfrac{47}{4}}$ is a complex number. Therefore, there are no real solutions.

Problem 16 : | *Media/Class Example*

Solve the following by completing the square. Be sure to simplify your answers. If answers are irrational, round to the nearest hundredth.

   a) $x^2 + x - 5 = 0$ $\qquad\qquad\qquad\qquad$ c) $3n^2 + 9n = 12$

   b) $2p^2 - p + 3 = 7$ $\qquad\qquad\qquad\qquad$ d) $y^2 = 8y - 20$

Problem 17 : | *You Try*

Solve the following by completing the square. Be sure to simplify your answers. If answers are irrational, round to the nearest thousandth.

   a) $n^2 + 3n = 2$ c) $4y^2 + 20y = -16$

   b) $3p^2 - 6p - 3 = 0$ d) $4x^2 + 4x + 17 = 5$

Practice Problems: *Completing the Square*

Find the constant term, c, to make each expression into a perfect square trinomial. Then factor the trinomial.

1. $a^2 + 4a + c$

2. $x^2 - 12x + c$

3. $m^2 - 6m + c$

4. $n^2 + 10n + c$

5. $y^2 - y + c$

6. $x^2 + 5x + c$

7. $r^2 - \frac{1}{3}r + c$

8. $g^2 + \frac{4}{5}g + c$

Solve each quadratic equation by completing the square. Be sure to simplify answers. If answers are irrational, round to the nearest hundredth.

9. $n^2 - 8n + 7 = 0$

10. $y^2 + 4y = 12$

11. $8g^2 + 16g = 64$

12. $x^2 - 8x - 12 = 0$

13. $3x^2 - 6x + 48 = 0$

14. $8b^2 + 16b - 37 = 5$

15. $m^2 = -15 + 9m$

16. $v^2 = 14v + 36$

17. $5k^2 - 10k - 45 = 0$

18. $x^2 + 16x + 55 = 5$

19. $2k^2 - 4k - 10 = -2$

20. $b^2 + 7b = 3$

21. $4a^2 + 16a - 1 = 0$

22. $5y^2 - 8y - 4 = 1$

23. $3p^2 = 2p + 6$

24. $3w^2 = 2 - w$

8.3. THE QUADRATIC FORMULA

Objective: Solve quadratic equations by using the quadratic formula.

Sometimes solving a quadratic equation by factoring or completing the square can be nearly impossible or quite cumbersome, especially when the leading coefficient of the quadratic expression is not 1. In those cases, it may be convenient to use the quadratic formula.

The quadratic formula is derived by completing the square of the quadratic equation, $ax^2 + bx + c = 0$. We will go through the derivation. The steps may look overwhelming but they are exactly the steps used in completing the square except for the trinomial coeefficients being non-numeric.

$$ax^2 + bx + c = 0 \qquad \text{Separate constant and variable terms by}$$
$$\text{subtracting } c \text{ on each side of equation}$$

$$ax^2 + bx = -c \qquad a \neq 1; \text{ Divide each term by } a$$

$$x^2 + \frac{b}{a}x = -\frac{c}{a} \qquad \text{Find } \left(\frac{1}{2} \cdot b\right)^2 = \left(\frac{1}{2} \cdot -\frac{b}{a}\right)^2 = \left(-\frac{b}{2a}\right)^2 = \frac{b^2}{4a^2}$$

$$\text{Add } \frac{b^2}{4a^2} \text{ to each side of equation}$$

$$x^2 + \frac{b}{a}x + \frac{b^2}{4a^2} = \frac{b^2}{4a^2} - \frac{c}{a} \qquad \text{Factor trinomial on left;}$$

$$\text{Find common denominator on right}$$

$$\left(x + \frac{b}{2a}\right)^2 = \frac{b^2}{4a^2} - \frac{4ac}{4a^2} \qquad \text{Combine constant terms}$$

$$\left(x + \frac{b}{2a}\right)^2 = \frac{b^2 - 4ac}{4a^2} \qquad \text{Take square root of each side}$$

$$\sqrt{\left(x + \frac{b}{2a}\right)^2} = \sqrt{\frac{b^2 - 4ac}{4a^2}} \qquad \text{Use Square Root Property}$$

$$\left|\left(x + \frac{b}{2a}\right)\right| = \frac{\sqrt{b^2 - 4ac}}{2a} \qquad \text{Solve absolute value equation}$$

$$x + \frac{b}{2a} = \pm\frac{\sqrt{b^2 - 4ac}}{2a} \qquad \text{Solve for } x$$

$$x = -\frac{b}{2a} \pm \frac{\sqrt{b^2 - 4ac}}{2a} \qquad \text{Rewrite solution as a single fraction}$$

$$x = \frac{-b \pm \sqrt{b^2 - 4ac}}{2a} \qquad \text{The Quadratic Formula}$$

This formula is important to us because we can use it to solve any quadratic equation. Once the quadratic equation is in standard form, $ax^2 + bx + c = 0$, identify what $a, b,$ and c are and substitute those values into $x = \dfrac{-b \pm \sqrt{b^2 - 4ac}}{2a}$ and we will get our solutions.

$$\text{Quadratic Formula: If } ax^2 + bx + c = 0 \text{ then } x = \frac{-b \pm \sqrt{b^2 - 4ac}}{2a}$$

Steps to using the quadratic formula when solving a quadratic equation:

1. Write the quadratic equation in standard form: $ax^2 + bx + c = 0$.

2. Identify the values of a (coefficient of x^2), b (coefficient of x) and c (the constant term).

3. Put the values of $a, b,$ and c into the quadratic formula.

4. Use order of operations to simplify solutions.

Problem 1 : | *Worked Example*

Solve $x^2 + 3x + 2 = 0$ by using the quadratic formula. Be sure to simplify your answers. If answers are irrational, round to the nearest hundredth.

Solution.

Let us go through each step in solving the quadratic equation by using the quadratic formula.

1. Equation already in standard form $\qquad\qquad x^2 + 3x + 2 = 0$

2. Identify the values of $a, b,$ and c $\qquad\qquad a = 1, b = 3, c = 2$

3. Put values into quadratic formula $\qquad x = \dfrac{-(3) \pm \sqrt{(3)^2 - 4(1)(2)}}{2(1)}$

4. Simplify $\qquad\qquad\qquad\qquad\qquad x = \dfrac{-3 \pm \sqrt{9 - 8}}{2}$

$$x = \frac{-3 \pm \sqrt{1}}{2}$$

Rational answers $\qquad\qquad\qquad x = \dfrac{-3 \pm 1}{2}$

Perform indicated operation $\qquad x = \dfrac{-3 + 1}{2}$ or $x = \dfrac{-3 - 1}{2}$

$$x = \frac{-2}{2} \text{ or } x = \frac{-4}{2}$$

Our Solutions $\qquad\qquad\qquad\quad x = -1$ or $x = -2$

Check to verify we have the correct solutions.

When $x = -1$:

$(-1)^2 + 3(-1) + 2 \overset{?}{=} 0$

$\qquad 1 - 3 + 2 = 0 \quad \checkmark$

When $x = -2$:

$(-2)^2 + 3(-2) + 2 \overset{?}{=} 0$

$\qquad 4 - 6 + 2 = 0 \quad \checkmark$

Note. The above example can also be solved by factoring or completing the square. All of them will yield the same answer. Verify it.

Problem 2 : | *Worked Example*

Solve $2y^2 = 2y + 1$ by using the quadratic formula. Be sure to simplify your answers. If answers are irrational, round to the nearest hundredth.

Solution.

1. Put equation in standard form;
   Subtract $2y$ and 1 from each side

   $$2y^2 - 2y - 1 = 0$$

2. Identify the values of $a, b,$ and c

   $$a = 2, b = -2, c = -1$$

3. Put values into quadratic formula

   $$y = \frac{-(-2) \pm \sqrt{(-2)^2 - 4(2)(-1)}}{2(2)}$$

4. Simplify

   $$y = \frac{2 \pm \sqrt{4+8}}{4}$$

   $$y = \frac{2 \pm \sqrt{12}}{4}$$

   Factor 2 from numerator

   $$y = \frac{2 \pm 2\sqrt{3}}{4}$$

   Simplify fraction

   $$y = \frac{2\left(1 \pm \sqrt{3}\right)}{4}$$

   Our Solutions are irrational

   $$y = \frac{1 \pm \sqrt{3}}{2}$$

   Solutions rounded to nearest hundredth $y \approx 1.37$ or -0.37

Problem 3 : | *Worked Example*

Solve $3m^2 + 4m = 4m - 2$ by using the quadratic formula. Be sure to simplify your answers. If answers are irrational, round to the nearest hundredth.

Solution.

1. Put equation in standard form;
   Subtract $4m$ and add 2 to each side

   $$3m^2 + 2 = 0$$

2. Identify the values of $a, b,$ and c

   $$a = 3, b = 0, c = 2$$

3. Put values into quadratic formula

   $$m = \frac{-(0) \pm \sqrt{(0)^2 - 4(3)(2)}}{2(3)}$$

4. Simplify

   $$m = \frac{0 \pm \sqrt{0 - 24}}{6}$$

   $$m = \frac{\pm\sqrt{-24}}{6}$$

$\sqrt{-24}$ is a complex number. Therefore, there are no real solutions.

Problem 4 : | *Worked Example*

Solve $12n - 4 = 9n^2$ by using the quadratic formula. Be sure to simplify your answers. If answers are irrational, round to the nearest hundredth.

Solution.

1. Put equation in standard form;
   Subtract $9n^2$ from each side $\qquad -9n^2 + 12n - 4 = 0$

2. Identify the values of $a, b,$ and c $\qquad a = -9, b = 12, c = -4$

3. Put values into quadratic formula $\quad n = \dfrac{-(12) \pm \sqrt{(12)^2 - 4(-9)(-4)}}{2(-9)}$

4. Simplify $\qquad\qquad\qquad\qquad n = \dfrac{-12 \pm \sqrt{144 - 144}}{-18}$

$$n = \dfrac{-12 \pm \sqrt{0}}{-18}$$

$$n = \dfrac{-12 \pm 0}{-18}$$

$$n = \dfrac{-12}{-18}$$

Our Solution $\qquad n = \dfrac{2}{3}$

Note. In solving quadratic equations, sometimes we get 2 distinct solutions, sometimes no real solutions and sometimes one unique solution, as seen in the above example.

Problem 5 : | *Media/Class Example*

Solve the following equations using the quadratic formula. Be sure to simplify your answers. If answers are irrational, also round to the nearest hundredth.

a) $x^2 + 6x + 8 = 0$

c) $\dfrac{2}{3}y^2 = \dfrac{4}{9}y + \dfrac{1}{3}$

b) $3n^2 - 4n = 5$

d) $0.5m^2 = 0.3$

Problem 6 : | *You Try*

Solve the following equations using the quadratic formula. Be sure to simplify your answers. If answers are irrational, also round to the nearest hundredth.

a) $x^2 = 3x + 5$

c) $5h - 3 = 6h^2$

b) $2n^2 - 7 = 0$

d) $\frac{2}{5}y^2 + \frac{2}{5}y + \frac{1}{10} = 0$

World View Note: This alternate derivation of the quadratic formula was known to the Hindus as early as AD 1025. It was still being taught as late as 1905 but was somehow replaced by the derivation done at the beginning of this section. This alternate derivation involves completing the square but does not require the leading coefficient to be 1. You will find that it is shorter, has simpler computations and does not involve fractions until the last step.

$$\text{Put equation in standard form} \qquad ax^2 + bx + c = 0$$

$$\text{Multiply each term by } 4a \qquad 4a^2x^2 + 4abx + 4ac = 0$$

$$\text{Separate variable and constant terms} \qquad 4ax^2 + 4abx = -4ac$$

$$\text{Add } b^2 \text{ to each side of equation} \qquad 4ax^2 + 4abx + b^2 = b^2 - 4ac$$

$$\text{Factor perfect square trinomial} \qquad (2ax + b)^2 = b^2 - 4ac$$

$$\text{Take square root of each side} \qquad \sqrt{(2ax + b)^2} = \sqrt{b^2 - 4ac}$$

$$\text{Apply square root property} \qquad |2ax + b| = \sqrt{b^2 - 4ac}$$

$$\text{Solve absolute value equation} \qquad 2ax + b = \pm\sqrt{b^2 - 4ac}$$

$$\text{Solve for } x; \text{ Subtract } b \text{ from each side} \qquad 2ax = -b \pm \sqrt{b^2 - 4ac}$$

$$\text{Divide each side by } 2a \qquad x = \frac{-b \pm \sqrt{b^2 - 4ac}}{2a}$$

We have the quadratic formula!

Practice Problems: *The Quadratic Formula*

Solve each equation using the quadratic formula and simplify your answers. If answers are irrational, round to the nearest hundredth.

1. $y^2 + 3y - 1 = 0$

2. $v^2 - 4v - 5 = -8$

3. $3p^2 - 5p + 3 = 4p^2$

4. $1 - \dfrac{2}{3}m^2 = 0$

5. $12g = 9g^2 + 4$

6. $y^2 - 4y = 1$

7. $3r^2 = 2r + 1$

8. $2x + 15 = 2x^2$

9. $m^2 - 14m + 55 = 0$

10. $k^2 = 3k + 5$

11. $2n^2 + 7n = 49$

12. $\dfrac{2}{5}b^2 = \dfrac{3}{5} - b$

13. $0.3n^2 - 1 = 0$

14. $r^2 + 4 = -6r$

15. $y^2 + \dfrac{1}{4} = -y$

16. $3v^2 = 2 + 3v$

17. $0 = 2x^2 + 5x + 3$

18. $1.6n^2 + 2.4n + 0.9 = 0$

8.4. STRATEGIES FOR SOLVING QUADRATIC EQUATIONS

Objective: To develop strategies for efficiently solving quadratic equations.

At this point, we have seen three approaches to solving quadratic equations:

- Factoring (if possible)

- Square Root Property or Completing the Square

- Quadratic Formula

Each technique has its advantages and limitations, and in this section, we seek to develop strategies to determine which method will lead to the most efficient solution. Let us solve a few equations using the three methods and then summarize what we find.

Problem 1 : | *Worked Example*

Solve the equation $x^2 + 2x - 35 = 0$ using the three different techniques. Then fill the table with pros and cons of each method and discuss your preference.

Solution.

a) **Factoring (if possible)**

$$x^2 + 2x - 35 = 0 \quad \text{Factor left side}$$
$$(x+7)(x-5) = 0 \quad \text{Use Zero-Factor Property}$$
$$x+7 = 0 \text{ or } x-5 = 0 \quad \text{Solve for } x$$
$$x = -7 \text{ or } x = 5 \quad \text{Our Solutions}$$

b) **Square Root Property or Completing the Square**

$$x^2 + 2x - 35 = 0 \quad \text{Separate constant and variable term by adding 35 to each side}$$
$$x^2 + 2x = 35 \quad \text{Find } \left(\frac{1}{2} \cdot b\right)^2 = \left(\frac{1}{2} \cdot 2\right)^2 = (1)^2 = 1; \text{ Add 1 to each side of equation}$$
$$x^2 + 2x + 1 = 35 + 1 \quad \text{Factor trinomial; add constat terms}$$
$$(x+1)^2 = 36 \quad \text{Take square root of each side}$$
$$\sqrt{(x+1)^2} = \sqrt{36} \quad \text{Apply square root property on left; simplify on right}$$
$$|x+1| = 6 \quad \text{Solve absolute value equation}$$
$$x+1 = \pm 6 \quad \text{Solve for } x$$
$$x = -1 \pm 6 \quad \text{Rational answers; Simplify solutions}$$
$$x = -1 + 6 \text{ or } x = -1 - 6 \quad \text{Perform indicated operation}$$
$$x = 5 \text{ or } x = -7 \quad \text{Our Solutions}$$

c) **Quadratic Formula**

For the equation $x^2 + 2x - 35 = 0$, identify $a = 1, b = 2$ and $c = -35$. We then apply the quadratic formula to find our solutions.

$$x = \frac{-b \pm \sqrt{b^2 - 4ac}}{2a}$$

$$= \frac{-(2) \pm \sqrt{(2)^2 - 4(1)(-35)}}{2(1)}$$

$$= \frac{-2 \pm \sqrt{4 + 140}}{2}$$

$$= \frac{-2 \pm \sqrt{144}}{2}$$

$$= \frac{-2 \pm 12}{2}$$

$$= \frac{-2 + 12}{2} \text{ or } \frac{-2 - 12}{2}$$

$$= \frac{10}{2} \text{ or } \frac{-14}{2}$$

$$= 5 \text{ or } -7$$

Technique	Pros	Cons
Factoring	Leads directly to solution with minimal step	
Square Root/ Completing the Square		Have to rewrite the equation to complete the square; There are a number of simplifying steps
Quadratic Formula		There are a number of simplifying steps

Problem 2 : | *Worked Example*

Solve the equation $(x-5)^2-4=3$ using the three different techniques. Then fill the table with pros and cons of each method and discuss your preference.

Solution.

a) **Factoring (if possible)**

$$(x-5)^2-4=3 \quad \text{Write the equation in standard form}$$
$$x^2-10x+25-4=3$$
$$x^2-10x+21=3 \quad \text{Subtract 3 on each side of the equation}$$
$$x^2-10x+18=0 \quad \text{Factor left side, if possible}$$

The left side of the equation is a prime polynomial, ie, there are no two integers whose product is 18 and whose sum is -10. Thus, we cannot solve this equation by factoring.

b) **Square Root Property or Completing the Square**

$$(x-5)^2-4=3 \quad \text{Isolate square term by adding 4 to each side}$$
$$(x-5)^2=7 \quad \text{Take square root of each side}$$
$$\sqrt{(x-5)^2}=\sqrt{7} \quad \text{Apply square root property on left}$$
$$|x-5|=\sqrt{7} \quad \text{Solve absolute value equation}$$
$$x-5=\pm\sqrt{7} \quad \text{Solve for } x$$
$$x=5\pm\sqrt{7} \quad \text{Our Solutions}$$

c) **Quadratic Formula**

Just like the factoring technique above, we need to first write the equation in standard form: $x^2-10x+18=0$. Identify $a=1, b=-10$ and $c=18$. We then apply the quadratic formula to find our solutions.

$$x = \frac{-b\pm\sqrt{b^2-4ac}}{2a}$$

$$= \frac{-(-10)\pm\sqrt{(-10)^2-4(1)(18)}}{2(1)}$$

$$= \frac{10\pm\sqrt{100-72}}{2}$$

$$= \frac{10\pm\sqrt{28}}{2}$$

$$= \frac{10 \pm 2\sqrt{7}}{2}$$

$$= \frac{2\left(5 \pm \sqrt{7}\right)}{2}$$

$$= 5 \pm \sqrt{7}$$

Technique	Pros	Cons
Factoring		Cannot be used
Square Root/ Completing the Square	Square term can be easily isolated and removed using the square root property; Most efficient way of finding solution	
Quadratic Formula		Have to write equation in standard form first; There are a number of simplifying steps

Problem 3 : | Media/Class Example

Solve the following equations using the three different techniques. Then make a table with pros and cons of each method and discuss your preference.

a) $x^2 - 6x = -5$ b) $x^2 - 6x = -4$

Problem 4 : | *Worked Example*

Solve the following equations using the three different techniques. Then make a table with pros and cons of each method and discuss your preference.

a) $x^2 = 3x$

c) $9y^2 = 32$

b) $x^2 = 3x + 1$

d) $-1 = 3 - (5 - 2y)^2$

SUMMARY: Now that we have worked through a number of examples, there are a few key observations we should keep in mind when solving a quadratic equation.

- Is there a **single square term** in the equation, e.g. $x^2 = 3$ or $3 + 2x^2 = 9$ or $(3x + 4)^2 + 2 = 8$? If so, then the **square root property** will typically lead to the most efficient solution.

- Is there **both a square term and a linear term** in the equation, e.g. $x^2 = 3x$ or $3x + 2x^2 = 9$ or $(3x + 4)^2 + 2x = 8$? If so, then you'll want to:

  i. Write the equation in standard form and try to solve the equation by **Factoring.** If the polynomial does not factor, then

  ii. Use either the **Quadratic Formula** or **Complete the Square.** Note that if the coefficient of the square term is not 1 and/or the resulting linear term is odd, then completing the square will be quite messy, and so the quadratic formula would be the ideal method to use.

Problem 5 : | *Media/Class Example*

Identify the most efficient technique to solve the given quadratic equation and indicate the main steps you would take in solving the equation. You **do not have to solve** the equations.

a) $4x^2 = 3x + 10$

c) $2p^2 + 3p = 4p + 5$

b) $7 - (y + 3)^2 = 1$

d) $w^2 + 4w = 4$

Problem 6 : | *You Try*

Solve each of the following equations using the method of your choice.

a) $3x - 5x^2 = -4$

e) $(3p - 1)^2 + 6p = 1$

b) $200y^2 = 3500 + 900y$

f) $c^2 - 8c = 4$

c) $4a^2 + 11 = 18$

g) $\frac{1}{2}(4q + 7)^2 = 12$

d) $10w^2 + 15w = 0$

Practice Problems: *Strategies for Solving Quadratic Equations*

1. Solve each equation using the indicated technique. Then state which technique you find most efficient.

    a) Factoring (if possible)

    b) Square Root Property/Completing the Square

    c) Quadratic Formula

      i. $x^2 - 6x + 8 = 0$ iv. $z^2 + 4z = 3$

      ii. $y^2 = 12y$ v. $(x-6)^2 - 3 = 9$

      iii. $9b^2 - 81 = 0$ vi. $2v^2 = 12 - 5v$

2. Identify the most efficient technique to solve the given quadratic equation and indicate the main steps you would take in practice to solve the equation. You don't have to solve the equation for this problem (though you may want to, for extra practice).

    a) $3y^2 - 8y - 35 = 0$ e) $A^2 + 10A = 2$

    b) $x(x+6) = 16$ f) $5 + (2b-7)^2 = 10$

    c) $3z^2 = 12$ g) $5 + (2b-7)^2 = 10b$

    d) $3z^2 = 12z$

3. Solve each equation using any method of your choice.

    a) $x(x-3) = 0$ j) $9c^2 + 4 = 12c$

    b) $x(x-3) = 4$ k) $y(3y-2) = 8$

    c) $3y^2 = 9y$ l) $x^2 + 6x = 9$

    d) $3y^2 = 9$ m) $-6B^2 - 12B + 50 = 4B - 20$

    e) $4x^2 + 4x - 3 = 0$ n) $-2x(2-x) = 8 + 4x$

    f) $(2x-7)^2 = 5$ o) $3w^2 - 4w + 11 = 0$

    g) $(2x-7)^2 = 0$ p) $\frac{1}{2}t^2 + 2t = 6$

    h) $6v^2 - v = 12$ q) $200x^2 - 300x = 400$

    i) $4b^2 - 8b - 7 = 0$ r) $\frac{3}{4}(b-5)^2 = 6$

8.5. Graphs of Quadratic Equations

Objective: To graph quadratic equations.

In this section we will learn how to graph quadratic equations, that is equations of the form $y = ax^2 + bx + c$.

The graph of a quadratic equation consists of all set of pairs (x, y) that make the equation true. For example, if we consider the equation $y = 3x^2 - 2x + 1$

- the point $(2, 9)$ is on the graph of the equation $y = 3x^2 - 2x + 1$, since
$$2 = 3(2)^2 - 2(2) + 1$$

- the point $(1, 1)$ is not in the graph of the equation $y = 3x^2 - 2x + 1$ since
$$1 \neq 3(1)^2 - 2(1) + 1$$

Problem 1 : | *Worked Example*

Find at least 5 points and use them to sketch the graph of the equation $y = x^2 + 1$

Solution.

Lets start by creating a table solutions for this equation.

x	pair	work
-2	$(-2, 5)$	$(-2)^2 + 1 = 5$
-1	$(-1, 2)$	$(-1)^2 + 1 = 2$
0	$(0, 1)$	$0^2 + 1 = 1$
1	$(1, 2)$	$1^2 + 1 = 2$
2	$(2, 5)$	$2^2 + 1 = 5$

Now we obtain our graph by plotting the points we found and connecting them with a smooth curve. The graph of a quadratic equation will not be a line as we can see from the points we obtained, the shape is called a parabola.

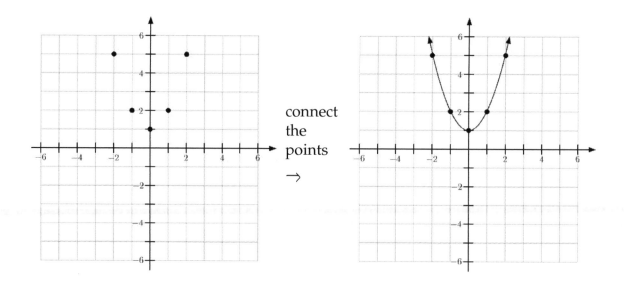

connect
the
points

$\rightarrow$

Problem 2 : | *In-class/Media Example*

Find at least 5 points and use them to sketch the graph of the equation $y = \dfrac{-1}{4}x^2 + 4$

Solution.

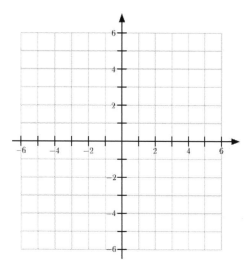

Problem 3 : | *You Try*

Find at least 5 points and use them to sketch the graph of the equation $y = \frac{1}{2}x^2 - 2$

Solution.

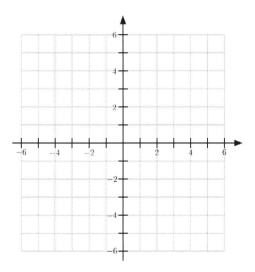

Problem 4 : | *In-class/Media Example*

Find at least 5 points and use them to sketch the graph of the equation $y = -\frac{x^2}{4} - \frac{x}{4} + \frac{3}{2}$

Solution.

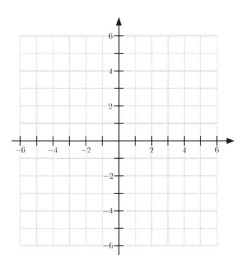

After we have worked with a few examples we can notice that the graph of a quadratic equation $y = ax^2 + bx + c$ is a curve (not a line), called a **parabola** that has the following properties.

- The graph can be opening up, or opening down (this depends on the sign of the coefficient a of the quadratic term).

- The graph can have a highest point (if it opens down), or a lowest point (if it opens down). This point is called the **vertex** of the parabola.

- The graph will always have a y-intercept.

- The graph can have up to 2 x-intercepts.

- The graph is symmetric about a vertical line that goes through the vetex. This line is called the axis of symmetry.

- If the graph has 2 x-intercepts, then the x-coordinate of the vertex is at the midpoint between the two x-intercepts.

From now on we will focus on finding these characteristics of a parabola, instead of finding a list of points on the graph.

Problem 5 : | *Worked Example*

Sketch the graph of the equation $y = x^2 + 2x - 8$. The graph should contain at least 5 points. including

a) the y-intercept

b) any x-intercepts

c) the vertex,

d) the line of symmetry.

Solution.

We first notice that since a is postive, the parabola opens up.

a) To find the y-intercept we make the substitution $x = 0$
$$y = 0 + 0 - 4$$
so the y-intercept is at $(0, -4)$.

b) To find the x-intercepts we make the substitution $y = 0$ and solve for x. In this case we can solve by factoring:
$$\begin{aligned} x^2 + 2x - 8 &= 0 \\ (x-2)(x+4) &= 0 \\ x = 2 \qquad &\text{or} \quad x = -4 \end{aligned}$$
so the x-intercepts are $(2, 0)$ and $(-4, 0)$.

c) As stated above, the x-coordinate of the vertex is at the midpoint between the two x-intercepts, that is, at $\dfrac{2 + (-4)}{2} = \dfrac{-2}{2} = -1$.

To find the y-coordinate of the vertex we substitute $x=-1$ into the equation
$$y = (-1)^2 + 2(-1) - 8$$
$$= 1 - 2 - 8$$
$$= -9$$
so the vertex is at $(-1, -9)$

d) the line of symmetry is a vertical line that goes through the vertex, so it is the line $x=-1$

We can now put all this together in a graph:

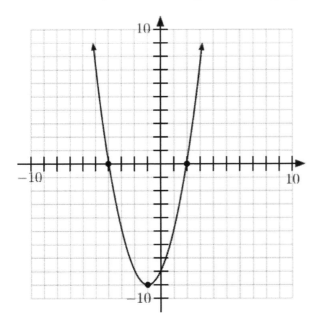

In the past example we found the x-intercepts by factoring. We can also find them using the quadratic formula:
$$x = \frac{-b \pm \sqrt{b^2 - 4ac}}{2a}$$

$$\frac{-b}{2a} \pm \frac{\sqrt{b^2 - 4ac}}{2a}$$

$$= \frac{-2}{2(1)} \pm \frac{\sqrt{4 - 4(-8)}}{2(1)}$$

$$= -1 \pm 3$$

We can read from this that the two x-intercepts will be 3 units away from -1, one on each side. The x-coordinate of the vertex will be precisely at $x = -1$.

> In general, for a quadratic equation $y = ax^2 + bx + c$ the x-coordinate of the vertex is equal to
> $$\frac{-b}{2a}$$

We obtain the y-coordinate of the vertex by substituting the x-coordinate in to the equation.

Problem 6 : | *Worked Example*

Sketch the graph of the equation $y = -x^2 - 2x + 3$. The graph should include

a) the $y-$intercept

b) any x-intercepts

c) the vertex

d) the line of symmetry

Solution.

a) To find the y-intercept we make the substitution $x = 0$
$$y = 0 + 0 + 3$$
so the y-intercept is at $(0, 3)$.

b) To find the x-intercepts we substitute $y = 0$ and solve by factoring:
$$\begin{aligned} -x^2 - 2x + 3 &= 0 \\ x^2 + 2x - 3 &= 0 \\ (x+3)(x-1) &= 0 \\ x = -3 \quad &\text{or} \quad x = 1 \end{aligned}$$

c) To find the x-coordinate of the vertex:
$$x = \frac{-b}{2a} = \frac{2}{2(-1)} = -1$$

We now substitute this value in the equation to obtain the y-coordinate of the vertex:
$$\begin{aligned} y &= -(-1)^2 - 2(-1) + 3 \\ &= -1 + 2 + 3 \\ &= 4 \end{aligned}$$
So the vertex is at $(-1, 4)$.

d) The line of symmemtry is $x = -1$

We can now put all this together in the graph:

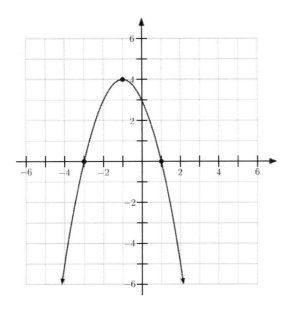

Problem 7 : | *Class/Media Example*

Sketch the graph of the equation $y = x^2 - 3x + 2$. The graph should include

a) the y-intercept

b) any x-intercepts

c) the vertex

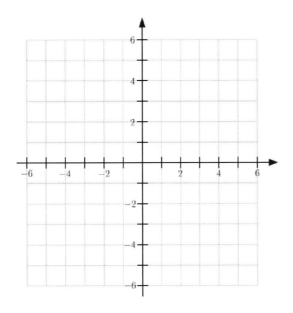

Problem 8 : | *You Try*

Sketch the graph of the equation $y = x^2 - 4x + 3$. The graph should include

    a) the y-intercept

    b) any x-intercepts

    c) the vertex

Solution.

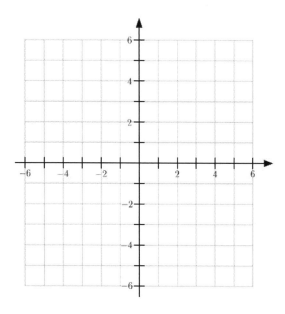

Problem 9 : | *Worked Example*

Sketch the graph of the equation $y = x^2 - 4x + 2$. The graph should include

    a) the y-intercept

    b) any x-intercepts

    c) the vertex

    d) the line of symmetry

Solution.

    a) The y-intercept is at $(0, 2)$.

    b) To find the x-intercepts we solve using the quadratic formula.

$$x = \frac{4 \pm \sqrt{16 - 4(2)}}{2}$$

$$= \frac{4}{2} \pm \frac{\sqrt{8}}{2}$$

$$= 2 \pm \frac{2\sqrt{2}}{2}$$

$$= 2 \pm \sqrt{2}$$

$$x \approx 0.6 \qquad x \approx 3.4$$

c) The x-coordinate of the vertex is at $x = \frac{4}{2} = 2$. We substitute this into the equation and obtain

$$y = 4 - 8 + 2$$
$$= 2$$

So the vertex is at $(2, -2)$

We can now sketch the graph:

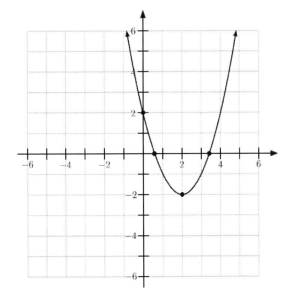

Problem 10 : | *Class/Media Example*

Sketch the graph of the equation $y = x^2 - 3x - 2$. The graph should include

a) the y-intercept

b) any x-intercepts

c) the vertex

d) the line of symmetry

Solution.

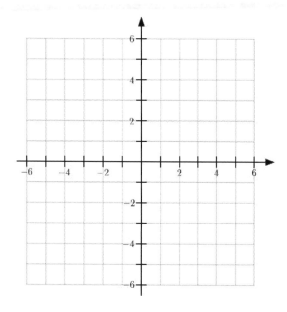

Problem 11 : | *You Try*

Sketch the graph of the equation $y = x^2 + 4x + 1$. The graph should include

   a) the y-intercept

   b) any x-intercepts

   c) the vertex

   d) the line of symmetry

Solution.

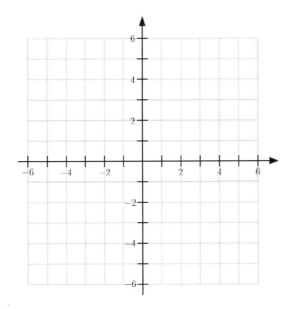

Problem 12 : | *Worked Example*

Sketch the graph of the equation $y = x^2 + 2$. The graph should include

    a) the y-intercept

    b) any x-intercepts

    c) the vertex

    d) the line of symmetry

Solution.

Since $a > 0$, we notice that the parabola opens up.

    a) The y-intercept is at $(0,2)$.

    b) To find the x-intercepts we set $y = 0$ and solve

$$\begin{aligned} x^2 + 2 &= 0 \\ x^2 &= -2 \\ |x| &= \sqrt{-2} \qquad \text{no solution} \end{aligned}$$

    since we found no solutions for the x-intercepts, this parabola does not intersect the x-axis.

    c) The x-coordinate of the vertex is at $\dfrac{-b}{2a} = \dfrac{-0}{2(1)} = 0$. So the vertex is at the point $(0,2)$, which is also the y-intercept.

    d) the line of symmetry is the the line $x = 0$, which is the y-axis.

Notice that in this case we don't have enough points to sketch a graph. We can obtain points as in Examples 1 and 2, by using different values for x. It is useful to find points symetrical around the vertex, so we use $x = -2$ and $x = 2$:

$$\begin{array}{ll} f(-2) = (-2)^2 + 2 & f(2) = (2)^2 + 2 \\ \quad\quad = 4 + 2 & \quad\quad = 4 + 2 \\ \quad\quad = 6 & \quad\quad = 6 \end{array}$$

The graph is pictured below:

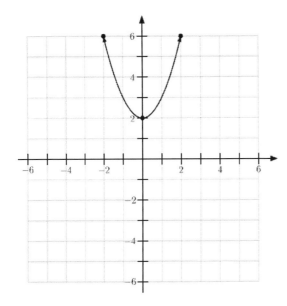

Problem 13 : | *Worked Example*

Sketch the graph of the equation $y = \dfrac{x^2}{2} + 3x + 5$. The graph should include

a) the y-intercept

b) any x-intercepts

c) the vertex

d) the line of symmetry

Solution.

a) The y-intercept is at $(0,5)$.

b) To find the x-intercepts we set $y = 0$ and solve using the quadratic formula.

$$x = \frac{-3 \pm \sqrt{9 - 4\left(\frac{1}{2}\right)(5)}}{2\left(\frac{1}{2}\right)}$$

$$= \quad -3 \pm \sqrt{-1} \qquad \text{No Solution}$$

Since we found no solutions for the x-intercepts, the graph does not intersect the x-axis,

c) The x-coordinate of the vertex is at $x = -3$. We substitute this into the equation and obtain

$$y = \frac{9}{2} + 3(-3) + 5$$

$$= \frac{9 - 18 + 10}{2}$$

$$= \frac{1}{2}$$

So the vertex is at $\left(-3, \frac{1}{2}\right)$

We can now sketch the graph:

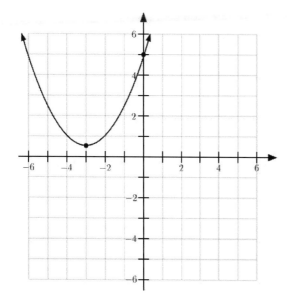

Practice Problems: *Graphing Quadratic Equations*

Sketch the graphs of each quadratic equation. Your graph must include

    a) the y-intercept

    b) any x-intercepts

    c) the vertex

    d) the line of symmetry

    1. $y = x^2 + 2x - 3$

    2. $y = -x^2 + 4x - 12$

    3. $y = -x^2 + 6x - 8$

    4. $y = x^2 + 4x + 4$

    5. $y = -x^2 - 4$

    6. $y = x^2 + 2x + 1$

    7. $y = \dfrac{x^2}{2} - x$

    8. $y = x^2 + 4x$

    9. $y = -x^2 + 6x$

    10. $y = x^2 - 5x + 6$

8.6. APPLICATIONS OF QUADRATICS- PART II

Objective: Solve application problems involving quadratic equations.

In application problems involving quadratic equations, questions regarding maximum or minimum usually involve determining the vertex. Remember that if a quadratic equation is in standard form, $y = ax^2 + bx + c$, the x-coordinate of the vertex is $x = -\dfrac{b}{2a}$. The y-coordinate of the vertex is obtained by substituting the x-value into the equation.

Problem 1 : | *Media/Class Example*

A local park facility has 100 feet of fencing and they want to use that to create a rectangular playing area. The amount of area, A, based on the width, x, is given by the following equation: $A = 50x - x^2$.

    a) What is the maximum area that can be surrounded with 100 feet of fencing?

    b) What is the width of the rectangular area?

Problem 2 : | *You Try*

A manufacturer has weekly production costs, given by the following equation: C = $0.25x^2 - 11x + 216$, where C is the total costs in dollars, and x is the number of units produced.

    a) What is the minimum weekly production cost?

    b) How many units can be produced at that cost?

| Problem 3 : | Media/Class Example |

A company earns a daily profit of P dollars by selling x items, modelled by the following equation: $P = -0.5x^2 + 40x - 300$.

    a) What is the daily profit if the company sells 25 items?

    b) How many items must the company sell in order to earn \$50 a day?

    c) How many items need to be sold to achieve maximum daily profit?

    d) What is the maximum daily profit?

| Problem 4 : | You Try |

A model rocket launched straight up from the ground achieves a height, h, in feet above the ground according to the equation $h = -16t^2 + 192t$, $t > 0$ where t is the time after the launch in second.

    a) How many seconds will it take until the rocket is 300 feet in the air? (Round to the nearest tenth)

    b) How high is the rocket after 6 seconds?

    c) When will the rocket hit the ground?

    d) What is the rocket's maximum height?

    e) How many seconds after launch will the rocket reach maximum height?

When application problems involve geometry, it is always a good idea to draw a picture first. This will help facilitate in setting up the equation.

Pythagorean Theorem

Given a right triangle with legs a and b, and hypotenuse, c, the Pythagorean Theorem states that the sum of the squares of the legs of a right triangle is equal to the square of its hypotenuse:

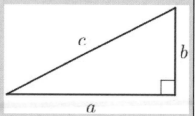

$$a^2 + b^2 = c^2.$$

Note. The hypotenuse of a right triangle is the longest side of the right triangle. It is the side that is opposite the right angle.

Problem 5 : | *Worked Example*

A rectangle is 10 inches long and 17 inches wide. What is the length of the diagonal of this rectangle? Give both exact answer and approximate answer rounded to one decimal place.

   Solution.

To solve this, we are going to use the Pythagorean Theorem. It helps to start with a diagram.

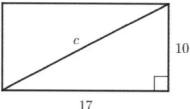

We can substitute 10 and 17 for a and b, and solve for c.

$$10^2 + 17^2 = c^2$$
$$100 + 289 = c^2$$
$$389 = c^2$$

At this point, we can use the Square Root Property to solve for c.

$$c = \pm\sqrt{389}$$

This does not simplify any further.

Given that we are looking for a length, $-\sqrt{389}$ is not a reasonable answer. So we can omit it.

The length of wire needed is $\sqrt{389}$ inches. This is our exact answer.

The length is approximately 19.7 inches.

Problem 6 : | *Media/Class Example*

A pole stands in a field and the owner wants to secure it with a guy-wire (a wire stretched from some place on the pole down to the ground, pulled away from the base). The owner has 40 feet of guy-wire. He wants to attach the wire to the pole at a certain height that is twice the distance the wire is from the base of the pole. Find how high above the ground is the guy-wire attached to the pole.

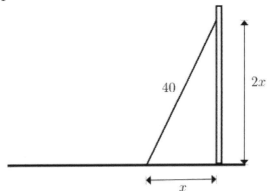

Problem 7 : | *You Try*

A right triangle has a hypotenuse of 25 mm. One side is 5 mm longer than the other side. Find the lengths of the two unknown sides.

Problem 8 : | *Media/Class Example*

The area of a triangle is 12 square feet. The height of the triangle is 5 feet more than its base. Find the dimensions of the triangle.

Problem 9 : | *You Try*

The area of a rectangle is 88 square meters. Its width is 3 meters shorter than its length. Find the dimensions of the rectangle.

Problem 10 : | *You Try*

If the area of a rectangle is 220.5 square feet and the perimeter is 63 feet, find the dimensions of the rectangle.

Problem 11 : | *Media/Class Example*

If the area of a rectangle is 104 square meters and the perimeter is 42 meters, find the dimensions of the rectangle.

Practice Problems: *Applications of Quadratics - Part II*

Solve each of the following application problem. Be sure to show all your work and include units in your answer.

1. The cost, C, of selling x algebra textbooks is $C = \frac{1}{4}x^2 - 35x + 2000$.

    a) How many algebra textbooks must be sold to minimize cost?

    b) What is the minimum cost?

2. A local coffee shop has determined that its daily revenue, R, is dependent on the price of a cup of coffee, p, based on the equation $R = -120p^2 + 684p$.

    a) If the price of a cup of coffee is $2.00, what is the daily revenue?

    b) If the coffee shop wants to earn $972 in daily revenue, how much should they charge for a cup of coffee?

    c) What is the coffee shop's maximum daily revenue?

    d) At what price per cup of coffee will maximum daily revenue be achieved?

3. The graph of the height, h, in feet, of a ball t seconds after it has been kicked is shown above. Using the graph, answer the following questions.

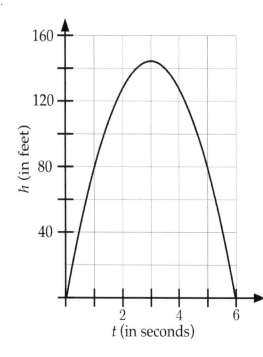

a) How hight did the ball get?
b) How many seconds did it take to reach that height?
c) How many seconds did it take for the ball to come back to the ground?
d) The equation of this graph is $h = -16t^2 + 96t$. Confirm your answers above algebraically.

4. A 52-inch TV has a diagonal of 52 inches. If the height of the TV is 29 inches, how wide is it? Round your answer to the nearest whole number.

5. The hypotenuse of a triangle is 7 inches long. One leg is 3 inches longer than the other. Find the length of each leg. Round answers to the nearest tenth.

6. The length of a rectangular garden is twice the width. The area of the garden is 80 square yards. Find the dimensions of the garden.

7. When each side of a square is increased by 4 inches, the area becomes 9 times larger than the original square. Find the length of the side of the original square.

8. The length of a rectangular garden is 3 yards longer than the width. The area of the garden is 54 square yards. Find the dimensions of the garden.

9. The infield of a baseball field is a diamond, with a base at each corner. Each corner is a right angle. The distance from home plate to 1st base is 90 ft., as is the distance from 1st to 2nd base, 2nd to 3rd base, and 3rd base to home.

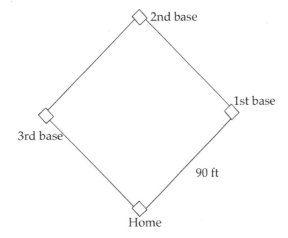

a) If a ball is thrown from home plate straight to 2nd base, how far will it travel? Give both exact and approximate answers rounded to two decimal places.

b) If the average time for a ball thrown from home plate straight to 2nd base is 2.0 seconds, at what speed was the ball thrown in feet/second? in miles/hour? Round your answers to the nearest whole number.

CHAPTER 8 ASSESSMENT

Simplify the following expressions, if possible.

1. $\sqrt{4x^2}$

2. $\sqrt{9-5y}$

Solve each of the following equation by the indicated methd. Give your answer in exact and simplified form. If the answer is irrational, round to the nearest hundredth.

3. Using the Square Root Property

    a) $x^2=20$
                b) $(3y-7)^2=25$

4. By Completing the Square

    a) $w^2+6w+4=0$
                b) $x^2-10x-11=0$

    c) $m^2-4m=16$
                 d) $3y^2-9y-12=0$

5. Using the Quadratic Formula

    a) $w^2-8w-3=0$
                 b) $2x^2-3x-2=0$

    c) $2m^2-4m=-1$
                 d) $\frac{1}{3}y^2+\frac{1}{6}y-\frac{1}{2}=0$

Sketch the graph of each quadratic equation. Your graph must include:

    a) y-intercept, written as an ordered pair

    b) any x-intercept(s), written as ordered pair(s)

    c) vertex

    d) axis of symmetry

    6. $y=x^2-4x+3$
                7. $y=-2x^2+6$

Solve the following word problems.

8. Jack wants to build a flower bed int he shape of an isosceles triangle. He wants the longest side of the triangle to be 34 feet. Find the dimensions of the flower bed. Round your answer to the nearest tenth and be sure to include units.

9. Jenny is designing a banner for the Founder's Day parade. The banner is 12 feet longer than it is wide and the total area is 64 square feet. Find the length and width of the banner.

CHAPTER 9

RADICALS PART II

9.1. SIMPLIFYING RADICALS

Objective: Simplify expressions involving square roots and cube roots.

Previously, we saw how to simplify numerical expressions with square roots. In this section, we will learn how to simplify algebraic expressions with different roots. First let us learn some vocabulary.

Given $\sqrt[n]{a}$, then:

- $\sqrt{}$ **is the radical sign**

- n **is the index**

- a **is the radicand**

The index must always be a positive integer greater than 1. When there is no index written, it is assumed to be 2 which means we have a square root.

Note. Do not confuse $\sqrt[n]{a}$ with $n\sqrt{a}$. They are two different expressions. The index of a radical expression is a superscript written on the upper left-hand corner of the radical sign. It is very important that the index is written properly.

It is also important to say radical expressions correctly.

Mathematically	*In Words*
$\sqrt{m}$	the square root of m
$\sqrt[3]{n}$	the cube root of n
$4\sqrt{x}$	Four times the square root of x
$5p^2\sqrt[3]{y}$	Five p squared times the cube root of y

How do radicals (or roots) work? Roots perform the inverse operation of applying positive integer exponents to a base. That is, $\sqrt[n]{a} = b$ because $b^n = a$. For example, $\sqrt{100} = 10$ because $10^2 = 100$.

Problem 1 : | *Worked Example*

Simplify the following.

a) $\sqrt{49}$

c) $\sqrt[3]{125}$

b) $\sqrt{-4}$

d) $\sqrt[3]{-125}$

Solution.

a) $\sqrt{49} = 7$. Think of a number whose square is 49. Since $7^2 = 49$, the square root of 49 must be 7.

b) $\sqrt{-4}$ is a complex number. There is no real number, such that when you multiply it by itself will give an answer -4.

c) $\sqrt[3]{125} = 5$. Think of a number whose cube is 125. Since $5^3 = 125$, the cube root of 125 must be 5.

d) $\sqrt[3]{-125} = -5$. Think of a number whose cube is -125. Since $(-5)^3 = -125$, the cube root of -125 must be -5.

Note. For square roots, a negative radicand yields no real number solution. For cube roots, a negative radicand yields a negative answer, that is because you can multiply a negative number an odd number of times and get a negative result.

Problem 2 : | *Media/Class Example*

Simplify the following.

a) $\sqrt{16}$

c) $\sqrt[3]{8}$

b) $\sqrt{-16}$

d) $\sqrt[3]{-8}$

Problem 3 : | *You Try*

Simplify the following.

a) $\sqrt{64}$

c) $\sqrt[3]{64}$

b) $-\sqrt{64}$

d) $\sqrt[3]{-64}$

We will not always encounter perfect roots. In simplifying radicals, take note of the index. If the index is 2, you want to find perfect square factors. If the index is 3, you want to find perfect cube factors. In some cases, we will also need to use the product rule for radicals to simplify further.

Product Rule for Radicals: $\sqrt[n]{ab} = \sqrt[n]{a} \cdot \sqrt[n]{b}$

(if n is even, $a > 0$ and $b > 0$; if n is odd, a and b can be any real number)

Problem 4 : | *Worked Example*

Simplify the following.

a) $\sqrt{72}$

b) $\sqrt[3]{72}$

Solution.

a)

$\sqrt{72}$ — Factor 72 such that there is a perfect square factor

$= \sqrt{36 \cdot 2}$ — Apply product rule for radicals

$= \sqrt{36} \cdot \sqrt{2}$ — Simplify $\sqrt{36}$

$= 6\sqrt{2}$ — Our Solution

b)

$\sqrt[3]{72}$ — Factor 72 such that there is a perfect cube factor

$= \sqrt[3]{8 \cdot 9}$ — Apply product rule for radicals

$= \sqrt[3]{8} \cdot \sqrt[3]{9}$ — Simplify $\sqrt[3]{8}$

$= 2\sqrt[3]{9}$ — Our Solution

Note. A radical expression is in its simplest form if no perfect root can be factored.

Problem 5 : | *Media/Class Example*

Simplify the following.

a) $\sqrt{24}$

b) $\sqrt[3]{24}$

Problem 6 : | *You Try*

Simplify the following.

a) $\sqrt{32}$

b) $\sqrt[3]{32}$

If there is a coefficient in front of the radical, the operation between the coefficient and radical is multiplication.

Problem 7 : | *Worked Example*

Simplify the following.

a) $4\sqrt{180}$ b) $-7\sqrt[3]{250}$

Solution.

a) $4\sqrt{180}$ Factor 180 such that there is a perfect square factor

     $= 4\sqrt{36\cdot 5}$ Apply product rule for radicals

     $= 4\sqrt{36}\cdot\sqrt{5}$ Simplify $\sqrt{36}$

     $= 4\cdot 6\cdot\sqrt{5}$ Multiply coefficients

     $= 24\sqrt{5}$ Our Solution

b) $-7\sqrt[3]{250}$ Factor 250 such that there is a perfect cube factor

     $= -7\sqrt[3]{125\cdot 2}$ Apply product rule for radicals

     $= -7\sqrt[3]{125}\cdot\sqrt[3]{2}$ Simplify $\sqrt[3]{125}$

     $= -7\cdot 5\cdot\sqrt[3]{2}$ Multiply coefficients

     $= -35\sqrt[3]{2}$ Our Solution

Problem 8 : | *Media/Class Example*

Simplify the following.

a) $5\sqrt{63}$ b) $4\sqrt[3]{-81}$

Problem 9 : | *You Try*

Simplify the following.

a) $-7\sqrt{12}$ b) $5\sqrt[3]{-54}$

Let's take a look at coefficients with fractions.

| **Problem 10 :** | **Worked Example** |

Simplify the following.

a) $\dfrac{3}{7}\sqrt{20}$

b) $\dfrac{\sqrt[3]{-16}}{4}$

Solution.

a) $\quad \dfrac{3}{7}\sqrt{20}$ $\qquad$ Factor 20 such that there is a perfect square factor

$\quad = \dfrac{3}{7}\sqrt{4\cdot 5}$ $\qquad$ Apply product rule for radicals

$\quad = \dfrac{3}{7}\sqrt{4}\cdot\sqrt{5}$ $\qquad$ Simplify $\sqrt{4}$

$\quad = \dfrac{3}{7}\cdot 2\cdot\sqrt{5}$ $\qquad$ Multiply coefficients

$\quad = \dfrac{6}{7}\sqrt{5}$ $\qquad$ Our Solution

Note. $\dfrac{6}{7}\sqrt{5}$ can also be written as $\dfrac{6\sqrt{5}}{7}$

b) $\quad \dfrac{\sqrt[3]{-16}}{4}$ $\qquad$ Factor -16 such that there is a perfect cube factor

$\quad = \dfrac{\sqrt[3]{-8\cdot 2}}{4}$ $\qquad$ Apply product rule for radicals

$\quad = \dfrac{\sqrt[3]{-8}\cdot\sqrt[3]{2}}{4}$ $\qquad$ Simplify $\sqrt[3]{-8}$

$\quad = \dfrac{-2\sqrt[3]{2}}{4}$ $\qquad$ Simplify fraction

$\quad = \dfrac{-\sqrt[3]{2}}{2}$ $\qquad$ Our Solution

Note. $\dfrac{-\sqrt[3]{2}}{2}$ can also be written as $-\dfrac{1}{2}\sqrt[3]{2}$

Problem 11 : | *Media/Class Example*

Simplify the following.

a) $\dfrac{\sqrt{8}}{6}$

b) $\dfrac{2}{3}\sqrt[3]{54}$

Problem 12 : | *You Try Example*

Simplify the following.

a) $\dfrac{5}{9}\sqrt{18}$

b) $\dfrac{\sqrt[3]{24}}{8}$

Variables are sometimes part of the radicand as well. A radical expression is in its simplest form if no variable in the radicand has a higher power than the index.

Since variables can take on positive or negative values, we have to be very careful when simplifying. Recall the square root property from the previous chapter which states that $\sqrt{x^2}=|x|$.

What about cube roots? What is $\sqrt[3]{x^3} = ?$ Let us take a look at the following:

$$\text{When } x=2: \quad \sqrt[3]{(2)^3} \;=\; \sqrt[3]{8} \;=\; 2$$
$$\text{When } x=-2: \quad \sqrt[3]{(-2)^3} \;=\; \sqrt[3]{-8} \;=\; -2$$

From the above examples, we observe that when x is a positive or negative real number, $\sqrt[3]{x^3}$ is equal to x itself. Generally, $\sqrt[3]{x^3}=x$.

| **Problem 13 :** | **Worked Example** |

Simplify the following.

a) $\sqrt{y^3}$

b) $\sqrt[3]{m^2}$

Solution.

a)

$$\sqrt{y^3}$$

Factor y^3 such that there is a perfect square factor

$$= \sqrt{y^2 \cdot y}$$

Apply product rule for radicals

$$= \sqrt{y^2} \cdot \sqrt{y}$$

Recall $\sqrt{y^2} = |y|$

$$= |y|\sqrt{y}$$

Our solution

b) The radical expression is already in its simplest form because the radicand's power, 2, is less than the index, 3.

| **Problem 14 :** | **Media/Class Example** |

Simplify the following.

a) $\sqrt{a^4}$

c) $\sqrt[3]{x^4}$

b) $-2\sqrt{c^3}$

d) $\sqrt[3]{27n^3}$

| **Problem 15 :** | **You Try** |

Simplify the following.

a) $\sqrt{36m}$

d) $\sqrt[3]{m^5}$

b) $5\sqrt{m^2}$

e) $\sqrt[3]{-27p^2}$

c) $-3\sqrt{16m^2}$

f) $-\sqrt[3]{64c^3}$

Practice Problems: *Simplifying Radicals*

Simplify the following, if possible.

1. $\sqrt{36}$

2. $\sqrt{-36}$

3. $\sqrt[3]{-125}$

4. $\sqrt{28}$

5. $\sqrt[3]{3000}$

6. $\sqrt{98}$

7. $\sqrt[3]{-72}$

8. $7\sqrt{63}$

9. $7\sqrt{-63}$

10. $-7\sqrt{63}$

11. $5\sqrt[3]{16}$

12. $5\sqrt[3]{-16}$

13. $-5\sqrt[3]{-16}$

14. $\dfrac{\sqrt{50}}{10}$

15. $\dfrac{5\sqrt{48}}{4}$

16. $\dfrac{2}{3}\sqrt{54}$

17. $\sqrt{64y^2}$

18. $\sqrt{100n^3}$

19. $-\sqrt{100k^4}$

20. $\sqrt[3]{-64y^3}$

21. $5\sqrt[3]{8h^4}$

22. $-\sqrt[3]{27n^5}$

23. $-7\sqrt{36x^2}$

24. $\sqrt{32p^4}$

25. $-\sqrt{50g^2}$

26. $\sqrt{20a^3}$

27. $\sqrt[3]{20a^3}$

9.2. Adding and Subtracting Radicals

Objective: To be able to identify like radicals and add and subtract them.

Adding and subtracting radicals is very similar to adding and subtracting polynomials. We need to have like terms. When radicals are in simplified form, like terms mean they have the same index and the same radicand. If there is a variable in front of the radical, they also have to be the same variables raised to the same power to be considered like terms.

Problem 1 : | *Worked Example*

Determine whether each pair of radical expressions are alike or not. Explain why.

a) $\sqrt{5x}, \sqrt{3x}$

b) $\sqrt{7y}, \sqrt[3]{7y}$

c) $5\sqrt{2}, -8\sqrt{2}$

d) $3x\sqrt{5}, 3y\sqrt{5}$

Solution.

a) Unlike expressions because the radicands are different

b) Unlike expressions because index are different

c) Like expressions because both terms have the same radicand and index

d) Unlike expressions because the variable in front of the radical are different

---
Problem 2 : | *Media/Class Example*

Determine whether each pair of radical expressions are alike or not. Explain why.

a) $\sqrt[3]{2x}, \sqrt{2x}$

b) $\sqrt{5}, \sqrt{10}$

c) $\sqrt{12}, \sqrt{3}$

d) $7x\sqrt{2}, 5x^2\sqrt{2}$

Problem 3 : | *You Try*

Determine whether each pair of radical expressions are alike or not. Explain why.

a) $5\sqrt{2}, -8\sqrt{3}$

b) $\sqrt{8}, -3\sqrt{2}$

c) $2\sqrt{y}, 2\sqrt[3]{y}$

d) $6x\sqrt{7}, -2x\sqrt{7}$

When adding and subtracting radical expressions, we keep the root part unchanged while adding and subtracting numerical coefficients, much the same way as combining like terms with polynomials. Let us take a look at the similarity.

Problem 4 : | *Worked Example*

Combine like terms given the following expressions.

a) polynomial expression: $5x + 3x - 2x$

   radical expression: $5\sqrt{11} + 3\sqrt{11} - 2\sqrt{11}$

b) polynomial expression: $7x + 5x^2$

   radical expression: $7\sqrt[3]{x} + 5\sqrt{x}$

Solution.

a) For the polynomial expression, since all the terms are alike, we add and subtract the coefficients while leaving the variable unchanged. Therefore, $5x + 3x - 2x = 6x$ For the radical expression, since all the terms are alike, we add and subtract the coefficients while leaving the root part unchanged. Therefore, $5\sqrt{11} + 3\sqrt{11} - 2\sqrt{11} = 6\sqrt{11}$. Notice the similarity with the polynomial example.

b) For the polynomial expression, since the variable exponents are different, the terms are considered to be unlike and therefore cannot be combined.
   For the radical expression, since the index of each term are different, the terms are considered to be unlike and therefore cannot be combined.

Problem 5 : | *Media/Class Example*

Perform the indicated operation and simplify answers.

a) $2\sqrt{5} + 3\sqrt{5}$

c) $8\sqrt[3]{7} - \sqrt[3]{7}$

b) $7\sqrt{6} - 9\sqrt{6} + \sqrt{6}$

d) $5\sqrt{2} - 4\sqrt[3]{2} - 7\sqrt{2} + \sqrt[3]{2}$

Problem 6 : | *You Try*

Perform the indicated operation and simplify answers.

a) $\sqrt{3}+2\sqrt{3}$
          d) $5\sqrt[3]{6}+\sqrt[3]{6}$

b) $\sqrt{10}-5\sqrt{10}$
          e) $4\sqrt{5}-7\sqrt[3]{5}$

c) $\sqrt{7}-5\sqrt{7}+8\sqrt{7}$
          f) $\sqrt{2}-3\sqrt{5}-\sqrt{2}+\sqrt{5}$

Sometimes radical expressions seem to have no like terms. However, if we simplify the expression first, we may find they do in fact have like terms.

Problem 7 : | *Worked Example*

Perform the indicated operation and simplify your answer.

a) $7\sqrt{20}+\sqrt{45}$
          b) $6\sqrt{18}-2\sqrt{98}+\sqrt{50}$

Solution.

a)
$$
\begin{aligned}
7\sqrt{20}+\sqrt{45} &= 7\sqrt{4\cdot5}+\sqrt{9\cdot5} &&\text{Simplify perfect squares}\\
&= 7\cdot2\cdot\sqrt{5}+3\cdot\sqrt{5} &&\text{Multiply coefficients}\\
&= 14\sqrt{5}+3\sqrt{5} &&\text{Combine like terms}\\
&= 17\sqrt{5} &&\text{Our Solution}
\end{aligned}
$$

b)
$$
\begin{aligned}
6\sqrt{18}-2\sqrt{98}+\sqrt{50} && \text{Factor radicands}\\
= 6\sqrt{9\cdot2}-2\sqrt{49\cdot2}+\sqrt{25\cdot2} && \text{Simplify perfect squares}\\
= 6\cdot3\cdot\sqrt{2}-2\cdot7\cdot\sqrt{2}+5\cdot\sqrt{2} && \text{Multiply coefficients}\\
= 18\sqrt{2}-14\sqrt{2}+5\sqrt{2} && \text{Combine like terms}\\
= 9\sqrt{2} && \text{Our Solution}
\end{aligned}
$$

Problem 8 : | *Media/Class Example*

Perform the indicated operation and simplify your answer.

a) $4\sqrt{32} - \sqrt{200}$

b) $5\sqrt{28} + 2\sqrt{7} - \sqrt{63}$

Problem 9 : | *You Try*

Perform the indicated operation and simplify your answer.

a) $\sqrt{54} + 3\sqrt{24}$

b) $\sqrt{27} - 7\sqrt{3} + 2\sqrt{12}$

Practice Problems: *Adding and Subtracting Radicals*

Simiplify

1. $4\sqrt{3}+\sqrt{3}$

2. $2\sqrt{7}-5\sqrt{7}$

3. $\sqrt[3]{6}-4\sqrt[3]{6}$

4. $2\sqrt{7}-\sqrt[3]{7}$

5. $5\sqrt{5}-\sqrt{5}-3\sqrt{5}$

6. $-2\sqrt{6}-\sqrt{3}-3\sqrt{6}$

7. $3\sqrt{10}-5\sqrt[3]{10}-2\sqrt{10}$

8. $-3\sqrt{2}+3\sqrt{5}+7\sqrt{2}-4\sqrt{5}$

9. $3\sqrt{6}+3\sqrt{7}+8\sqrt{6}-2\sqrt{7}$

10. $\sqrt{11}+2\sqrt[3]{11}-2\sqrt{11}-5\sqrt[3]{11}$

11. $-7\sqrt[3]{6}+2\sqrt{6}+7\sqrt[3]{6}-\sqrt{6}$

12. $-\sqrt{2}+3\sqrt{8}$

13. $\sqrt{12}+3\sqrt{27}$

14. $\sqrt{54}-4\sqrt{6}$

15. $2\sqrt{2}-3\sqrt{18}-\sqrt{2}$

16. $-3\sqrt{27}+2\sqrt{3}-\sqrt{12}$

17. $-3\sqrt{18}-\sqrt{8}+2\sqrt{32}$

18. $6\sqrt{5}-\sqrt{20}-\sqrt{45}$

19. $3\sqrt{6}-3\sqrt{24}+\sqrt{54}$

20. $3\sqrt{8}-\sqrt{12}+\sqrt{48}$

21. $3\sqrt{54}-\sqrt{50}+2\sqrt{6}+2\sqrt{8}$

22. $\sqrt{90}-\sqrt{40}-\sqrt{80}+\sqrt{20}$

9.3. MULTIPLYING AND DIVIDING RADICALS

Objective: Multiply and divide radicals using the product and quotient rules for radicals.

Multiplication of Radicals

Multiplication of radicals can only be performed if the index on the all radicals are the same. If they are, the operation is performed by multiplying coefficients together and multiplying radicands together. This leads us to the more generalized form of the product rule for radicals.

Product Rule for Radicals: $\left(c\sqrt[n]{a}\right)\cdot\left(d\sqrt[n]{b}\right)=cd\sqrt[n]{ab}$

(if n is even, $a>0$ and $b>0$; if n is odd, a and b can be any real number)

Problem 1 : | *Worked Example*

Multiply the following radical expressions and simplify your answers.

a) $\sqrt{2}\cdot\sqrt{3}$

c) $\sqrt{20y}\cdot\sqrt{5y}$

b) $\left(-5\sqrt{10}\right)\cdot\left(4\sqrt{6}\right)$

d) $\left(5\sqrt[3]{4}\right)\cdot\left(\sqrt[3]{4}\right)$

Solution.

a) Multiply the coefficients. The result cannot be simplified further.

$$\sqrt{2}\cdot\sqrt{3} = \sqrt{6}$$

b) Multiply the coefficients together and then multiply the radicands together. Then simplify the result.

$$\begin{aligned}
\left(-5\sqrt{10}\right)\cdot\left(4\sqrt{6}\right) &= -5\cdot4\sqrt{10\cdot6}\\
&= -20\sqrt{60}\\
&= -20\sqrt{4\cdot15}\\
&= -20\cdot2\cdot\sqrt{15}\\
&= -40\sqrt{15}
\end{aligned}$$

c) Multiply the coefficients together and then simplify the result. Recall that $\sqrt{x^2} = |x|$.

$$\begin{aligned}
\sqrt{20y}\cdot\sqrt{5y} &= \sqrt{100y^2}\\
&= 10|y|
\end{aligned}$$

d) Multiply the coefficients together and then multiply the radicands together. Then simplify the result.

$$\left(5\sqrt[3]{4}\right)\cdot\left(\sqrt[3]{4}\right) = 5\sqrt[3]{4\cdot4}$$
$$= 5\sqrt[3]{16}$$
$$= 5\sqrt[3]{8\cdot2}$$
$$= 5\sqrt[3]{8}\cdot\sqrt[3]{2}$$
$$= 5\cdot2\cdot\sqrt[3]{2}$$
$$= 10\sqrt[3]{2}$$

Problem 2 : | *Media/Class Example*

Multiply the following radical expressions and simplify your answers.

a) $\left(\sqrt{7}\right)\cdot\left(\sqrt{14}\right)$

c) $\left(\sqrt{3n}\right)\cdot\left(\sqrt{12n^2}\right)$

b) $\left(7\sqrt{6}\right)\cdot\left(2\sqrt{12}\right)$

d) $\left(\sqrt[3]{9}\right)\cdot\left(\sqrt[3]{6}\right)$

Problem 3 : | *You Try*

Multiply the following radical expressions and simplify your answers.

a) $\left(\sqrt{12}\right)\cdot\left(\sqrt{3}\right)$

c) $\left(\sqrt{8a}\right)\cdot\left(\sqrt{2a}\right)$

b) $\left(-3\sqrt{5}\right)\cdot\left(\sqrt{10}\right)$

d) $\left(\sqrt[3]{3}\right)\cdot\left(\sqrt[3]{18}\right)$

Sometimes, the distributive property is used to multiply radicals.

Problem 4 : | *Worked Example*

Perform the indicated operation and simplify your answers.

a) $\sqrt{6}\,(3\sqrt{10}-\sqrt{15})$

b) $(6+\sqrt{7})\,(3-2\sqrt{7})$

Solution.

a) Use the distributive property to multiply. Then simplify the result.

$$\sqrt{6}\,(3\sqrt{10}-\sqrt{15}) = 3\sqrt{6\cdot10}-\sqrt{6\cdot15}$$
$$= 3\sqrt{60}-\sqrt{90}$$
$$= 3\sqrt{4\cdot15}-\sqrt{9\cdot10}$$
$$= 3\sqrt{4}\sqrt{15}-\sqrt{9}\sqrt{10}$$
$$= 3\cdot2\cdot\sqrt{15}-3\sqrt{10}$$
$$= 6\sqrt{15}-3\sqrt{10}$$

b) Use the distributive property to multiply. Then simplify the result.

$$\left(6+\sqrt{7}\right)\left(3-2\sqrt{7}\right) = 6\cdot3-6\cdot2\cdot\sqrt{7}+3\sqrt{7}-2\sqrt{7\cdot7}$$
$$= 18-12\sqrt{7}+3\sqrt{7}-2\sqrt{49}$$
$$= 18-9\sqrt{7}-2\cdot7$$
$$= 20-9\sqrt{7}-14$$
$$= 6-9\sqrt{7}$$

Problem 5 : | *Media/Class Example*

Perform the indicated operation and simplify your answers.

a) $\sqrt{5}\left(4-\sqrt{10}\right)$

c) $\left(2+\sqrt{6}\right)\left(2-\sqrt{6}\right)$

b) $2\sqrt{8}\left(5\sqrt{2}+\sqrt{3}\right)$

d) $(5-2\sqrt{3})\,(4+\sqrt{6})$

Problem 6 : | *You Try*

Perform the indicated operation and simplify your answers.

a) $2\sqrt{6}\left(3+\sqrt{2}\right)$

c) $\left(\sqrt{7}-\sqrt{3}\right)\left(\sqrt{7}+\sqrt{3}\right)$

b) $\sqrt{10}\left(7\sqrt{2}-\sqrt{3}\right)$

d) $\left(5+\sqrt{3}\right)^2$

Problem 7 : | *Media/Class Example*

Check if $x=1-2\sqrt{5}$ is a solution to the equation $x^2-2x=19$.

Problem 8 : | *You Try*

Check if $x=2+\sqrt{3}$ is a solution to the equation $x^2-4x+1=0$.

Division of Radicals

A radical is considered not simplified if it there is a fraction underneath the radical sign or if there is a radical in the denominator. To simplify, we will use a more generalized form of the quotient rule for radicals.

> **Quotient Rule of Radicals:** $\dfrac{a\sqrt[m]{b}}{c\sqrt[m]{d}} = \dfrac{a}{c}\sqrt[m]{\dfrac{b}{d}}$, where $c \neq 0$ and $d \neq 0$
>
> (if m is even, $b > 0$ and $d > 0$; if n is odd, b and d can be any real number, $d \neq 0$)

Problem 9 : | Worked Example

Simplify $\dfrac{\sqrt{48}}{\sqrt{27}}$

Solution.

$$\dfrac{\sqrt{48}}{\sqrt{27}} = \sqrt{\dfrac{48}{27}} \qquad \text{Use Quotient Rule then simplify fraction}$$

$$= \sqrt{\dfrac{16}{9}} \qquad \text{Use Quotient Rule going in the other direction}$$

$$= \dfrac{\sqrt{16}}{\sqrt{9}} \qquad \text{Simplify}$$

$$= \dfrac{4}{3} \qquad \text{Our Solution}$$

Problem 10 : | Worked Example

Simplify $11\sqrt[3]{\dfrac{54}{2}}$

Solution.

$$11\sqrt[3]{\dfrac{54}{2}} = 11\sqrt[3]{\dfrac{54}{2}}$$

$$= 11\sqrt[3]{27}$$

$$= 11 \cdot 3$$

$$= 33$$

Problem 11 : | *Media/Class Example*

Simplify the following.

a) $\dfrac{\sqrt{45}}{\sqrt{20}}$

c) $\dfrac{10\sqrt[3]{81}}{\sqrt[3]{24}}$

b) $\dfrac{9}{6}\sqrt{\dfrac{50}{8}}$

d) $\dfrac{1}{5}\sqrt[3]{-\dfrac{192}{3}}$

Problem 12 : | *You Try*

Simplify the following.

a) $12\sqrt{\dfrac{25}{9}}$

c) $\dfrac{\sqrt[3]{54}}{\sqrt[3]{-16}}$

b) $\dfrac{4\sqrt{75}}{15\sqrt{3}}$

d) $\dfrac{5}{8}\sqrt[3]{\dfrac{40}{5}}$

In most math texts, it is customary to clear the denominator of any radicals. The process is called rationalizing the denominator. Here, we will only focus on clearing the square root term in the denominator. If the radical in the denominator is a monomial and is not a perfect square, we rationalize the denominator by multiplying both numerator and denominator by the same square root term.

Problem 13 : | *Worked Example*

Simplify $\dfrac{2}{\sqrt{5}}$

Solution.

To simplify the expression, we will rationalize the denominator by multiplying both numerator and denominator by $\sqrt{5}$.

$$\begin{aligned}
\frac{2}{\sqrt{5}} &= \frac{2}{\sqrt{5}} \cdot \frac{\sqrt{5}}{\sqrt{5}} \\
&= \frac{2\sqrt{5}}{\sqrt{25}} \\
&= \frac{2\sqrt{5}}{5}
\end{aligned}$$

Problem 14 : | *Worked Example*

Simplify $\dfrac{\sqrt{15}}{\sqrt{6}}$

Solution.

We see that both radicals have the same index and the radicands both have a common factor 3. Instead of rationalizing the denominator right away, we can simplify first by applying the quotient rule for radicals.

$$\frac{\sqrt{15}}{\sqrt{6}} = \sqrt{\frac{15}{6}} \qquad \text{Apply Quotient Rule and simplify fraction}$$

$$= \sqrt{\frac{5}{2}} \qquad \text{Apply Quotient Rule going in the other direction}$$

$$= \frac{\sqrt{5}}{\sqrt{2}} \qquad \text{Rationalize denominator}$$

$$= \frac{\sqrt{5}}{\sqrt{2}} \cdot \frac{\sqrt{2}}{\sqrt{2}} \qquad \text{Multiply numerator and denominator by } \sqrt{2}$$

$$= \frac{\sqrt{10}}{\sqrt{4}} \qquad \text{Simplify denominator}$$

$$= \frac{\sqrt{10}}{2} \qquad \text{Our Solution}$$

Problem 15 : | *Class/Media Example*

Simplify each of the following.

a) $\dfrac{3}{\sqrt{2}}$

c) $\dfrac{\sqrt{10}}{\sqrt{14}}$

b) $\dfrac{9}{2\sqrt{3}}$

d) $\sqrt{\dfrac{11}{50}}$

Problem 16 : | *You Try*

Simplify each of the following.

a) $\sqrt{\dfrac{16}{5}}$

c) $\dfrac{\sqrt{12}}{\sqrt{18}}$

b) $\sqrt{\dfrac{1}{2}}$

d) $\dfrac{4\sqrt{7}}{\sqrt{8}}$

Practice Problems: *Multiplying and Dividing Radicals*

Perform the indicated operation and simplify.

1. $\sqrt{5} \cdot \sqrt{3}$

2. $-2\sqrt{10} \cdot \sqrt{5}$

3. $\left(9\sqrt[3]{-4}\right) \cdot \left(2\sqrt[3]{6}\right)$

4. $\left(3\sqrt{5}\right) \cdot \left(4\sqrt{6}\right)$

5. $\left(-4\sqrt[3]{9}\right) \cdot \left(\sqrt[3]{6}\right)$

6. $\sqrt{12m} \cdot \sqrt{3m}$

7. $\sqrt{5r} \cdot \sqrt{20r^2}$

8. $\sqrt{2y} \cdot \sqrt{8y}$

9. $\sqrt{6}(\sqrt{2}+2)$

10. $\sqrt{10}(\sqrt{5}+\sqrt{2})$

11. $5\sqrt{3}(3+\sqrt{6})$

12. $(2+2\sqrt{2})(3+\sqrt{2})$

13. $(2-\sqrt{3})(5+2\sqrt{3})$

14. $\left(4-\sqrt{7}\right)\left(4+\sqrt{7}\right)$

15. $\left(2\sqrt{6}+1\right)^2$

16. $\left(\sqrt{11}+\sqrt{6}\right)\left(\sqrt{11}-\sqrt{6}\right)$

17. $\left(3-\sqrt{7}\right)^2$

18. $(2\sqrt{3}+\sqrt{5})(5\sqrt{3}+4)$

19. $\dfrac{\sqrt{12}}{\sqrt{3}}$

20. $\dfrac{4\sqrt{15}}{\sqrt{3}}$

21. $\dfrac{4\sqrt{125}}{\sqrt{5}}$

22. $\dfrac{\sqrt{12}}{5\sqrt{100}}$

23. $\sqrt[3]{\dfrac{15}{64}}$

24. $\dfrac{3\sqrt[3]{10}}{5\sqrt[3]{27}}$

25. $\dfrac{2}{\sqrt{3}}$

26. $\dfrac{\sqrt{10}}{\sqrt{6}}$

27. $\sqrt{\dfrac{1}{6}}$

28. $\dfrac{\sqrt{2}}{3\sqrt{5}}$

Verify that x is a solution to the given equation.

29. $x=-5\sqrt{2}$; Equation: $x^2-50=0$

30. $x=2-\sqrt{2}$; Equation: $x^2-4x+2=0$

31. $x=-3+\sqrt{11}$; Equation: $x^2+6x=2$

9.4. Square root Equations

Objective: Solve a square root equation and check for extraneous solutions.

Consider an equation with a variable under the square root, such as $\sqrt{x} = 4$. We call this a square root equation. To solve this, we need to isolate the radical.

Recall from Chapter 7, $\left(\sqrt{b}\right)^2 = b$, for $b \geq 0$. Therefore, to solve $\sqrt{x} = 4$, we must square each side of the equation to keep it balanced.

Problem 1 : | *Worked Example*

Solve $\sqrt{x} = 3$ for x.

> **Solution.**
>
> $$\begin{aligned} \sqrt{x} &= 3 && \text{Square each side of equation} \\ (\sqrt{x})^2 &= (3)^2 && \text{Simplify each side} \\ x &= 9 && \text{Our Solution} \end{aligned}$$

Check the answer.
Substitute $x = 9$ back into the *original* equation to verify answer.

$$\sqrt{9} = 3 \quad \checkmark$$

Problem 2 : | *Worked Example*

Solve $\sqrt{y-4} = 5$ for y.

> **Solution.**
>
> $$\begin{aligned} \sqrt{y-4} &= 5 && \text{Square each side of equation} \\ (\sqrt{y-4})^2 &= (5)^2 && \text{Simplify each side} \\ y-4 &= 25 && \text{Solve for } x; \text{ This is a linear equation} \\ y &= 29 && \text{Our Solution} \end{aligned}$$

Check the answer.
Substitute $y = 29$ into the *original* equation to verify answer.

$$\sqrt{(29)-4} \overset{?}{=} 5$$
$$\sqrt{25} = 5 \quad \checkmark$$

Problem 3 : | *Worked Example*

Solve $\sqrt{2n} - 1 = 7$ for n.

    Solution.

Before we square each side, we must first isolate the radical.

$$\sqrt{2n} - 1 = 7 \qquad \text{Add 1 to each side}$$
$$\sqrt{2n} = 8 \qquad \text{Square each side}$$
$$\left(\sqrt{2n}\right)^2 = (8)^2 \qquad \text{Simplify}$$
$$2n = 64 \qquad \text{Solve for } x; \text{ This is a linear equation}$$
$$n = 32 \qquad \text{Our Solution}$$

Check the answer.

Substitute $n = 32$ into the *original* equation to verify answer.

$$\sqrt{2(32)} - 1 \overset{?}{=} 7$$
$$\sqrt{64} - 1 \overset{?}{=} 7$$
$$8 - 1 = 7 \qquad \checkmark$$

Problem 4 : | *Worked Example*

Solve $\sqrt{x+2} = -6$ for x.

    Solution.

Looking closely at the problem, you may notice that it has no solution. Why? Because the positive square root of a number can never be negative.

However, if you do not recognize that right away, you can always discover this *when you check your answer.*

The radical is isolated, so we can square both sides.

$$\left(\sqrt{x+2}\right)^2 = (-6)^2 \quad \text{Simplify}$$
$$x + 2 = 36 \qquad \text{Solve for } x; \text{ This is a lineqar equation}$$
$$x = 34 \qquad \text{Our Solution}$$

Check the answer.

Substitute $x = 34$ into the *original* equation to verify answer.

$$\sqrt{(34)+2} \overset{?}{=} -6$$
$$\sqrt{36} \neq -6$$

$x = 34$ does not solve the equation. Therefore, the equation has no solution. We say $x = 34$ is an *extraneous solution.*

Remark. An *extraneous solution* is an answer that emerges from the process of solving the problem, but is not a valid solution. When solving square root equations, it is very important to always verify that the answer is a valid solution to the given problem.

In summary, to solve a square root equation:

1. Isolate the radical expression

2. Square each side

3. Solve for the variable

4. Check your solution

Problem 5 : | *Media/Class Example*

Solve the following equations.

a) $\sqrt{x} - 1 = 7$

b) $-\sqrt{5k + 10} = -4$

c) $5 + \sqrt{x - 1} = 3$

Problem 6 : | *You Try*

Solve the following equations.

a) $\sqrt{p} = 9$

c) $\sqrt{x + 8} + 10 = 2$

b) $-3\sqrt{5w} = -30$

d) $\dfrac{\sqrt{6x + 4}}{3} = 2$

In solving square root equations, we sometimes encounter the square of a binomial. To expand the square of a binomial, remember to always rewrite the expression as a product of two binomials and then multiply by applying the distributive property. Let us review how to expand the square a binomial such as $(x-3)^2$.

$$
\begin{aligned}
(x-3)^2 &= (x-3)(x-3) \\
&= x^2 - 3x - 3x + 9 \\
&= x^2 - 6x + 9
\end{aligned}
$$

Problem 7 : | *Worked Example*

Solve $x = 5 + \sqrt{4x+1}$ for x.

   Solution.

x	$=$	$5 + \sqrt{4x+1}$	Isolate the radical
$x - 5$	$=$	$\sqrt{4x+1}$	Square each side
$(x-5)^2$	$=$	$\left(\sqrt{4x+1}\right)^2$	Simplify
$x^2 - 10x + 25$	$=$	$4x + 1$	This is a quadratic equation; set equal to 0
$x^2 - 14x + 24$	$=$	0	Factor left-hand side trinomial
$(x-12)(x-2)$	$=$	0	Solve for x

$$
\begin{aligned}
x - 12 = 0 \quad &\text{or} \quad x - 2 = 0 \\
x = 12 \quad &\text{or} \quad x = 2
\end{aligned}
$$

Check by substituting each value into the *original* equation *one at a time.*
When $x = 12$:

$$
\begin{aligned}
12 &\overset{?}{=} 5 + \sqrt{4(12)+1} \\
12 &\overset{?}{=} 5 + \sqrt{48+1} \\
12 &\overset{?}{=} 5 + \sqrt{49} \\
12 &= 5 + 7 \quad \checkmark
\end{aligned}
$$

When $x = 2$:

$$
\begin{aligned}
2 &\overset{?}{=} 5 + \sqrt{4(2)+1} \\
2 &\overset{?}{=} 5 + \sqrt{8+1} \\
2 &\overset{?}{=} 5 + \sqrt{9} \\
2 &\neq 5 + 3
\end{aligned}
$$

$x = 2$ is an extraneous solution. Therefore, $x = 12$ is our solution to the equation.

Problem 8 : | *Media/Class Example*

Solve the following equations.

a) $\sqrt{6n+7}=n+2$

b) $y+\sqrt{10-y}=4$

Problem 9 : | *You Try*

Solve the following equations.

a) $x=\sqrt{x+2}$

c) $\sqrt{x+11}=x-1$

b) $\sqrt{y}+2=y$

d) $2+\sqrt{1-8p}=p$

Problem 10 : | *Worked Example*

Solve $\sqrt{3x-8} - \sqrt{x} = 0$ for x.

Solution.

$$
\begin{array}{rcll}
\sqrt{3x-8} - \sqrt{x} &=& 0 & \text{Isolate the radicals} \\
\sqrt{3x-8} &=& \sqrt{x} & \text{Square each side} \\
\left(\sqrt{3x-8}\right)^2 &=& (\sqrt{x})^2 & \text{Simplify} \\
3x-8 &=& x & \text{Solve for } x;\ \text{Subtract } 3x \text{ on each side} \\
-8 &=& -2x & \text{Divide each side by } -2 \\
4 &=& x & \text{Our Solution}
\end{array}
$$

Check by substituting $x=4$ into the *original* equation.

$$
\begin{array}{rcl}
\sqrt{3(4)-8} - \sqrt{4} &\overset{?}{=}& 0 \\
\sqrt{12-8} - \sqrt{4} &\overset{?}{=}& 0 \\
\sqrt{4} - \sqrt{4} &=& 0 \qquad \checkmark
\end{array}
$$

Problem 11 : | *Media/Class Example*

Solve $\sqrt{x} = \sqrt{5x-4}$ for x.

Problem 12 : | *You Try*

Solve the following $\sqrt{3x+8} = \sqrt{4x+2}$ for x.

Problem 13 : | *Media/Class Example*

A simple pendulum consists of a string, cord or wire that allows a suspended mass to swing back and forth. The categorization of "simple" comes from the fact that all of the mass of the pendulum is concentrated in its "bob" or suspended mass. If the bob is pulled to the right and then released, the time for it to swing to the left and then back to its original positiion, known as the period, can be found with the following formula:

$T = 2\pi\sqrt{\dfrac{l}{g}}$ where T is the period in seconds, l is the length of the string in meters, and $g = 9.81\,m/\sec^2$ is the Earth's gravity.

a) If the bob of a pendulum is suspended from a one meter string, what is its period? Round your answer to the nearest integer.

b) If you want to double the period, how long must the string be? Round your naswer to the nearest integer.

Problem 14 : | *You Try*

Given two points (x_1, y_1) and (x_2, y_2), the distance, d, between them is given by the formula: $d = \sqrt{(x_2 - x_1)^2 + (y_2 - y_1)^2}$.

a) What is the distance between the points $(5, -3)$ and $(7, 1)$? Give the exact answer in simplified form. If the answer is irrational, provide an approximation rounded to the nearest tenth.

b) What is the distance between the points $\left(4\sqrt{2}, -\sqrt{3}\right)$ and $\left(3\sqrt{2}, \sqrt{3}\right)$? Give the exact answer in simplified form. If the answer is irrational, provide an approximation rounded to the nearest tenth.

c) Challenge: Find 2 points on the x-axis that are 5 units away from the point $(6, 3)$.

| Problem 15 : | *You Try* |

The formula $s = k\sqrt{d}$ relates the speed, s (in mph) of a car and the distance, d (in feet) of the skid when a driver hits the brakes. k is a constant that differs for different road conditions.

a) What was the speed of a car, if the skid is 124 feet long, on an icy pavement where $k = 25$?

b) If a car is traveling at 55 mph on wet pavement where $k = 3.24$, how far will the car skid? Round to the nearest foot.

Practice: *Square Root Equations*

Solve.

1. $\sqrt{x-4}=11$

2. $\sqrt{10-y}+2=8$

3. $10+\sqrt{5x+1}=3$

4. $\sqrt{7p+2}=4$

5. $\sqrt{6w+1}-7=0$

6. $-2\sqrt{3c+1}+15=9$

7. $\sqrt{6x-5}-x=0$

8. $2x=\sqrt{9x-2}$

9. $\sqrt{y}+2=y$

10. $n=1+\sqrt{7-n}$

11. $3+m=\sqrt{6m+13}$

12. $\sqrt{3-3g}=2g+1$

13. $h+\sqrt{4h+1}=5$

14. $\sqrt{x+2}-x=2$

15. $\sqrt{3k+22}=\sqrt{14-k}$

16. $\sqrt{1-3p}-\sqrt{3p+7}=0$

17. $\sqrt{3m}=\sqrt{\dfrac{m}{3}+8}$

Word Problems.

18. Find the distance between the following points. Be sure to simplify your answer. If answer is irrational, approximate to the nearest hundredth.

  a) $(0,7)$ and $(-3,1)$

  b) $\left(\sqrt{5},-\sqrt{2}\right)$ and $\left(-\sqrt{5},\sqrt{2}\right)$

19. A car is traveling 70 mph on dry pavement. Use the formula $s=k\sqrt{d}$, where s is the car speed (in mph), d is the distance (in feet) of the skid when a driver hits the brakes and $k=5.34$, to find the how far the car skids. Round to the nearest foot.

Challenge Problems.

20. Find 2 points on the y-axis that are 13 units away from the point $(-5,10)$.

21. Solve the following equations.

  a) $\sqrt{y+9}=3+\sqrt{y}$

  b) $\sqrt{c+1}=1+\sqrt{c-6}$

Rescue Roody.

Help Roody figure out his misunderstanding. Roody was asked to solve the following equation: $\sqrt{x+4}-2=x$. Why was Roody's answer marked incorrect?

$$\sqrt{x+4}-2 = x$$
$$\sqrt{x+4} = x+2$$
$$\left(\sqrt{x+4}\right) = (x+2)^2$$
$$x+4 = x^2+4$$
$$0 = x^2-x$$
$$0 = x(x-1)$$
$$x=0 \text{ or } x=1$$

Check.

When $x=0$:
$$\sqrt{(0)+4}-2 \overset{?}{=} 0$$
$$\sqrt{4}-2 = 0 \quad\checkmark$$

When x=1:
$$\sqrt{(1)+4}-2 \overset{?}{=} 1$$
$$\sqrt{5}-2 \neq 1$$

Therefore, $x=0$ is the solution to the equation.

CHAPTER 9 ASSESSMENT

Simplify the following.

1. $\sqrt{121}$

2. $\sqrt[3]{\dfrac{8}{125}}$

3. $-\sqrt{20}$

4. $\sqrt[3]{-80}$

5. $\sqrt{100y^6}$

6. $\sqrt[3]{64y^6}$

Perform the indicated operation and simplify your answer.

7. $10\sqrt{3} - 8\sqrt{3}$

8. $\sqrt[3]{10} + 4\sqrt[3]{10}$

9. $4\sqrt{3} + 5\sqrt{27}$

10. $6\sqrt{18} - \sqrt{50}$

11. $7\sqrt{5} + 2\sqrt{5} - 9\sqrt{5}$

12. $\sqrt{12} - 5\sqrt{48} - \sqrt{72}$

13. $\sqrt{6} \cdot \sqrt{3}$

14. $\left(2\sqrt{10}\right)\left(\sqrt{6}\right)$

15. $\sqrt{2}\left(3 + \sqrt{10}\right)$

16. $\left(5 + \sqrt{7}\right)\left(5 - \sqrt{7}\right)$

17. $\left(4 - \sqrt{6}\right)^2$

18. $\left(2\sqrt{3} + 1\right)\left(\sqrt{3} - 4\right)$

19. $\dfrac{5\sqrt{27}}{\sqrt{12}}$

20. $\sqrt{\dfrac{1}{7}}$

21. $\dfrac{4}{\sqrt{6}}$

22. $\dfrac{\sqrt{2}}{4\sqrt{5}}$

23. Verify that $x = 4 - \sqrt{30}$ is a solution to the equation: $x^2 - 8x = 14$

Solve the following equations.

24. $\sqrt{x+5} = 7$

25. $\sqrt{2a+5} + 8 = 3$

26. $\sqrt{n+20} = n$

27. $p = \sqrt{p}$

28. $\sqrt{3k-4} = \sqrt{k+3}$

29. $\sqrt{y-4} = y - 6$

CHAPTER 10

RATIONAL EXPRESSIONS

10.1. INTRODUCTION TO RATIONAL EXPRESSIONS

Objective: Evaluate and simplify rational expression.

Rational Expressions

Recall that a **rational number (or fraction)** is a real number that can be expressed as a ratio $\frac{p}{q}$, where p and q are integers and $q \neq 0$. Examples of rational numbers are:

$$\frac{2}{3}, \quad -\frac{9}{5}, \quad 7, \quad \text{and} \quad 0$$

Generalizing this notion, we define a **rational expression** to be an algebraic expression that can be expressed as a ratio $\frac{P}{Q}$, where P and Q are *polynomials*; this expression will be defined when $Q \neq 0$. Some examples of rational expressions are:

$$\frac{2x+3}{x-1}, \quad -\frac{3}{y^2}, \quad z^3, \quad 7, \quad \text{and} \quad 0$$

Note that every rational number is in fact a rational expression.

Rational Expressions: Ratios of Polynomials

Ex. $\dfrac{4x^2}{x+8}, \quad -\dfrac{11}{7y}, \quad 8z^2, \ldots$

Rational Numbers: Ratios of Integers

Ex. $\dfrac{4}{9}, \quad -\dfrac{11}{9}, \quad 8, \ldots$

This is a very important observation and forms the basis of the following theme we will see throughout this chapter.

IMPORTANT NOTE: When working with Rational Expressions, we use the same principles as when working with Fractions.

427

Evaluating Rational Expressions

Just as we evaluate other types of algebraic expressions such as polynomials, radical expressions, etc., we can evaluate rational expressions.

Problem 1 :	*Worked Example*

Evaluate the rational expression $\dfrac{3x^2-8}{x-2}$ for:

a) $x=3$

b) $x=-4$

c) $x=2$

    Solution.

For each part, we substitute the given value of x and simplify the resulting expression using the order of operation.

    a) When $x=3$:

$$\frac{3(3)^2-8}{(3)-2} = \frac{3(3)^2-8}{(3)-2}$$
$$= \frac{3(9)-8}{(3)-2}$$
$$= \frac{27-8}{1}$$
$$= 19$$

    b) When $x=-4$:

$$\frac{3(-4)^2-8}{(-4)-2} = \frac{3(16)-8}{(-4)-2}$$
$$= \frac{48-8}{(-4)-2}$$
$$= \frac{40}{-6}$$
$$= \frac{20}{-3}$$

    Note. Since $\dfrac{-a}{b}=\dfrac{a}{-b}=-\dfrac{a}{b}$, we can also write the solution as $-\dfrac{20}{3}$ or $\dfrac{-20}{3}$.

c) When $x = 2$:

$$\frac{3(2)^2 - 8}{(2) - 2} = \frac{3(4) - 8}{(2) - 2}$$
$$= \frac{12 - 8}{(2) - 2}$$
$$= \frac{4}{0}$$

Since it is not permissible to divide by 0, we conclude that the expression $\dfrac{3x^2 - 8}{x - 2}$ is **undefined** when $x = 2$.

Problem 2 : | *Media/Class Example*

Evaluate the expression $\dfrac{x + 1}{x^2 - 6x + 8}$ for:

a) $x = 1$

c) $x = -1$

b) $x = 0$

d) $x = 4$

Problem 3 : | *You Try*

Evaluate the expression $\dfrac{3 - y}{y^2 - 9}$ for:

a) $y = -2$

c) $y = -3$

b) $y = 0$

d) $y = 3$

Determining When a Rational Expression is Undefined

As we saw in the last two examples, a rational expression is undefined for values of the variable that make the denominator equal to 0. Thus, to find all values for which a rational expression is undefined, we simply set the denominator equal to 0 and solve the resulting equation.

Problem 4 : | Worked Example

For what values of the variable is the expression $\dfrac{x+3}{x^2-2x}$ undefined?

    Solution.

We set the denominator equation to 0 and solve the resulting equation.

$$\begin{aligned} x^2-2x &= 0 \quad \text{Factor the left-side} \\ x(x-2) &= 0 \quad \text{Use zero-product property} \\ x = 0 \quad &\text{or} \quad x-2 = 0 \\ & \qquad\quad x = 2 \end{aligned}$$

We conclude that the expression $\dfrac{x+3}{x^2-2x}$ is undefined when $x=0$ or $x=2$.

Problem 5 : | Media/Class Example

For what values of the variable is the given expression undefined?

a) $\dfrac{3x}{x-7}$

c) $\dfrac{4y}{y^2+5y+6}$

c) $\dfrac{1-p}{2p}$

d) $\dfrac{1}{n^2+1}$

Problem 6 : | You Try

For what values of the variable is the given expression undefined?

a) $\dfrac{2-x}{x-2}$

c) $\dfrac{z+4}{z^2+16}$

c) $\dfrac{5+y}{y^2}$

d) $\dfrac{3t-1}{2t^2-5t-3}$

Simplifying Rational Expressions

Let us recall how we simplify rational numbers (or fractions) by considering the fraction $\frac{8}{12}$. We first look for a common factor between 8 and 12 (4 in this case) and then simplify the fraction using division.

$$\frac{8}{12} = \frac{2 \cdot 4}{3 \cdot 4} = \frac{2}{3}$$

As we noted in the beginning of this section, a rational expression is really just a more general type of fraction. Thus, to simplify a rational expression, we use the exact same reasoning as in the above example.

To simplify a rational expression, we seek common factors between the numerator and denominator and then simplify the expression using division. In other words, we simplify rational expressions using the following two steps:

1. **Factor** both numerator and denominator completely.

2. Simplify the fraction by dividing common factors.

Problem 7 : | *Worked Example*

Simplify the given rational expressions.

a) $\dfrac{4x}{x^2}$

b) $\dfrac{2n+6}{n^2-9}$

c) $\dfrac{p^2-2p}{p-2}$

d) $\dfrac{y+3}{3+y}$

e) $\dfrac{z+7}{z}$

f) $\dfrac{2m^2}{5m-10}$

Solution.

a)

$$\frac{4x}{x^2} = \frac{4 \cdot x}{x \cdot x}$$

$$= \frac{4 \cdot \overset{1}{\cancel{x}}}{x \cdot \underset{1}{\cancel{x}}}$$

$$= \frac{4}{x}$$

b)

$$\frac{2n+6}{n^2-9} = \frac{2(n+3)}{(n-3)(n+3)}$$

$$= \frac{2(\cancel{n+3})^{1}}{(n-3)(\cancel{n+3})_{1}}$$

$$= \frac{2}{n-3}$$

c)

$$\frac{p^2-2p}{p-2} = \frac{p(p-2)}{(p-2)_{1}}$$

$$= \frac{p(\cancel{p-2})}{(\cancel{p-2})_{1}}$$

$$= \frac{p}{1}$$

$$= p$$

d)

$$\frac{y+3}{3+y} = \frac{y+3}{y+3}$$

$$= 1$$

e) For the expression $\frac{z+7}{z}$, the numerator and denominator are fully factored and there is no common factor amongst the numerator and denominator, so the expression is already simplified.

f)

$$\frac{2m^2}{5m-10} = \frac{2m^2}{5(m-2)}$$

Since there is no common factor amongst the numerator and denominator, the expression $\frac{2m^2}{5m-10}$ is in fact already in simplified form.

Note. For a problem such as f), it is acceptable to give your answer as either $\frac{2m^2}{5m-10}$ or the equivalent expression in factored form, $\frac{2m^2}{5(m-2)}$.

Warning. When simplifying rational expressions, it is crucial that you **first factor** both numerator and denominator completely and then simplify the expression by dividing *common factors*. For example, consider the expression $\dfrac{x^2}{3x+7}$. If we attempt to simplify this expression by dividing individual *terms* in the numerator and denominator, we run into problems such as this:

$$\frac{x^2}{3x+7} \neq \frac{x}{3+7} = \frac{x}{10}$$

To see why these are not equivalent expressions, let us consider the value $x=2$. For the expression $\dfrac{x^2}{3x+7}$, when $x=2$, we have $\dfrac{(2)^2}{3(2)+7} = \dfrac{4}{6+7} = \dfrac{4}{13}$.

Whereas, for the expression $\dfrac{x}{10}$, when $x=2$, we have $\dfrac{2}{10} = \dfrac{1}{5}$.

The mistake in dividing the x^2-term in the numerator by the x-term in the denominator is that x **is not a factor of the denominator!**

The factors of the numerator, x^2 are x and x, whereas the denominator has only the single factor $3x + 7$. Thus, there are no common factors between the numerator and denominator.

So the expression, $\dfrac{x^2}{3x+7}$ is already in simplified form.

Problem 8 : | *Media/Class Example*

Simplify the given rational expressions.

a) $\dfrac{3a}{6a^3}$

b) $\dfrac{(m-4)}{(m-4)^2}$

c) $\dfrac{6+w}{w+6}$

d) $\dfrac{x^2-4}{x^2+16}$

e) $\dfrac{y^2-3y}{2y}$

f) $\dfrac{n-3}{n^2+5n+6}$

Problem 9 : | *You Try*

Simplify the given rational expressions.

a) $\dfrac{2w^4}{6w^2}$

d) $\dfrac{v}{v+8}$

b) $\dfrac{9(z+7)}{6(z+7)^2}$

e) $\dfrac{-n-4}{n+4}$

c) $\dfrac{t+3}{t^2+6t+9}$

f) $\dfrac{y^2-5y}{y^2-25}$

The theme of the next examples is that the numerator and denominator share factors that are negatives of one another. The key to simplifying such expressions is to use the fact that

$$b-a = -(a-b)$$

Problem 10 : | *Worked Examples*

Simplify the given rational expressions.

a) $\dfrac{2-y}{y-2}$ b) $\dfrac{8-x}{x^2-8x}$ c) $-\dfrac{w-4}{4-w}$

    Solution.

a)

$$\frac{2-y}{y-2} = \frac{-(y-2)}{(y-2)_1}$$

$$= \frac{-(\cancel{y-2})}{(\cancel{y-2})_1}$$

$$= \frac{-1}{1}$$

$$= -1$$

Note. We could also have rewritten the denominator, $y - 2$ as $-(2 - y)$ and then simplified. The answer will be the same.

$$\frac{2-y}{y-2} = \frac{2-y}{-(2-y)}$$

$$= \frac{(2\!\!\!\not/\,y)}{-(2\!\!\!\not/\,y)}$$

$$= \frac{1}{-1}$$

$$= -1$$

b)

$$\frac{8-x}{x^2-8x} = \frac{-(x-8)}{x(x-8)}$$

$$= \frac{-(x\!\!\!\not/\,8)}{x(x\!\!\!\not/\,8)}$$

$$= \frac{-1}{x} \quad \text{or} \quad -\frac{1}{x}$$

c)

$$-\frac{w-4}{4-w} = -\frac{-(4-w)}{4-w}$$

$$= -\frac{-(4\!\!\!\not/\,w)}{(4\!\!\!\not/\,w)}$$

$$= -\frac{-1}{1}$$

$$= 1$$

Problem 11 : | *Media/Class Example*

Simplify the given rational expressions.

a) $\dfrac{3-g}{g-3}$

b) $\dfrac{4x-3}{6-8x}$

c) $-\dfrac{1-a}{a-1}$

Problem 12 : | *Media/Class Example*

Simplify the given rational expressions.

a) $\dfrac{5-m}{m-5}$

c) $\dfrac{y-7}{y^2-7y}$

b) $\dfrac{3x+6}{2+x}$

d) $-\dfrac{z-9}{9-z}$

Practice Problems: *Introduction to Rational Expressions*

Evaluate the expression for the given value. Simplify your answers.

1. $\dfrac{v+2}{v}$ when $v=4$

2. $\dfrac{3-b}{b-9}$ when $b=-3$

3. $\dfrac{x^2-3}{x^2-5x}$ when $x=-1$

4. $\dfrac{4-a^2}{a+4}$ when $a=-2$

5. $\dfrac{b+2}{b^2+b-12}$ when $b=3$

6. $\dfrac{n^2-n-6}{n-3}$ when $n=0$

Find all values of the variable that make the given rational expression undefined.

7. $\dfrac{k}{k+10}$

8. $\dfrac{n^2}{10n+5}$

9. $\dfrac{m+7}{3}$

10. $\dfrac{m^2-16}{2m}$

11. $\dfrac{r^2+2}{r^2+1}$

12. $\dfrac{x-1}{1-x}$

13. $\dfrac{2}{p^2-6p}$

14. $\dfrac{3n^2+6n}{n^2-9}$

15. $\dfrac{4a}{a^2-5a-6}$

16. $\dfrac{y-2}{3y^2-y-10}$

Simplify the given rational expression.

17. $\dfrac{4n}{12n^2}$

18. $\dfrac{32x^3}{8x^5}$

19. $\dfrac{h+5}{5+h}$

20. $\dfrac{a-7}{a+7}$

21. $\dfrac{20}{4p+2}$

22. $\dfrac{r-10}{r}$

23. $-\dfrac{m-3}{3-m}$

24. $\dfrac{8m+16}{2+m}$

25. $\dfrac{n-9}{81-9n}$

26. $\dfrac{x+1}{x^2+8x+7}$

27. $\dfrac{4x^2+28x}{4x^2}$

28. $\dfrac{-5b-20}{5b+20}$

29. $\dfrac{n^2+2n+1}{6+6n}$

30. $\dfrac{3a-10}{10+3a}$

31. $\dfrac{54+9v}{v^2-4v-60}$

32. $\dfrac{b^2+14b+48}{b^2+15b+56}$

33. $\dfrac{k^2-12k+32}{64-k^2}$

34. $\dfrac{x^2-5x+4}{3x^2-7x+4}$

Rescue Roody!

Roody was aksed to simplify the following rational expressions:

a) $\dfrac{a-4}{4a-16}$

b) $\dfrac{y+8}{2y}$

His work is shown below. All were marked incorrect. Help him understand what went wrong.

a) $\dfrac{a-4}{4a-16} = \dfrac{a-4}{4(a-4)} = \dfrac{a\cancel{-}4}{4(a\cancel{-}4)} = 4$

b) $\dfrac{y+8}{2y} = \dfrac{\cancel{y}+8}{2\cancel{y}} = \dfrac{8}{2} = 4$

10.2. Multiplying and Dividing Rational Expressions

Objective: To multiply and divide rational expressions.

As we saw in the last section, rational expressions are really just special types of fractions. Thus, to multiply and divide rational expressions, we must use the same techniques as when we multiply and divide fractions. Let us review how to do this.

Mutiplication of Fractions

To multiply two fractions, we use the property $\frac{a}{b} \cdot \frac{c}{d} = \frac{ac}{bd}$. However, we may be able to simplify first before multiplying, making the process easier. Here are some examples to illustrate.

Problem 1 :	*Worked Example*

Perform the indicated operation and provide your final answer in simplified form.

a) $\dfrac{5}{7} \cdot \dfrac{9}{4}$

b) $\dfrac{13}{49} \cdot \dfrac{14}{11}$

Solution.

a) There are no common factors between 5, 4, 7, and 9, so we cannot simplify. Multiply across.

$$\frac{5}{7} \cdot \frac{9}{4} = \frac{5 \cdot 9}{7 \cdot 4}$$
$$= \frac{45}{28}$$

Note. The improper fraction $\dfrac{45}{28}$ is in simplified form. There is no need to rewrite it as a mixed number.

b) 14 and 49 share a common factor of 7. Simplify first before multipying.

$$\frac{13}{49} \cdot \frac{14}{11} = \frac{13}{\underset{7}{\cancel{49}}} \cdot \frac{\overset{2}{\cancel{14}}}{11}$$
$$= \frac{13}{7} \cdot \frac{2}{11}$$
$$= \frac{26}{77}$$

Simplifying first before multiplying made the fraction multiplication much easier. We also arrived at an already simplified fraction for our final result. We will make use of the same idea when we multiply rational expressions.

Division of Fractions

To divide two fractions, we first convert the division problem into a multiplication problem: $\frac{a}{b} \div \frac{c}{d} = \frac{a}{b} \cdot \frac{d}{c}$. That is, to divide to fractions, we multiply the first fraction by the *reciprocal* of the second fraction.

| Problem 2 : | Worked Example |

Perform the indicated operation and provide your final answer in simplified form.

a) $\quad \dfrac{35}{9} \div 25$

b) $\quad \dfrac{\dfrac{10}{63}}{\dfrac{25}{18}}$

Solution.

a) First rewrite 25 as $\dfrac{25}{1}$ then multiply the first fraction by the reciprocal of the second fraction. Remember to simplify first before multiplying.

$$\frac{35}{9} \div 25 = \frac{35}{9} \div \frac{25}{1}$$
$$= \frac{35}{9} \cdot \frac{1}{25}$$
$$= \frac{\overset{7}{\cancel{35}}}{9} \cdot \frac{1}{\underset{5}{\cancel{25}}}$$
$$= \frac{7}{9} \cdot \frac{1}{5}$$
$$= \frac{7}{45}$$

b) The problem is the same as writing $\dfrac{10}{63} \div \dfrac{25}{18}$. Multiply $\dfrac{10}{63}$ by the *reciprocal* of $\dfrac{25}{18}$, that is $\dfrac{10}{63} \cdot \dfrac{18}{25}$. Note that 10 and 25 share a common factor of 5 while 18 and 63 share a common factor of 9. Simplify first before multiplying.

$$\frac{10}{63} \div \frac{25}{18} = \frac{10}{63} \cdot \frac{18}{25}$$
$$= \frac{\overset{2}{\cancel{10}}}{\underset{7}{\cancel{63}}} \cdot \frac{\overset{2}{\cancel{18}}}{\underset{5}{\cancel{25}}}$$
$$= \frac{2}{7} \cdot \frac{2}{5}$$
$$= \frac{4}{35}$$

Problem 3 : | *You Try*

Perform the indicated operation and provide your final answer in simplified form.

a) $\dfrac{99}{82} \cdot \dfrac{41}{66}$

c) $-\dfrac{4}{7} \div \dfrac{-21}{2}$

c) $-144 \cdot \dfrac{13}{24}$

d) $\dfrac{\frac{81}{5}}{9}$

Rational Expressions

We are now ready to turn to rational expressions. We will use the same techniques as above. However, it is crucial that we **factor** the numerators and denominators completely before proceeding.

Problem 4 : | *Worked Example*

Perform the indicated operation and provide your final answer in simplified form.

a) $\dfrac{4}{z^2} \cdot \dfrac{z}{2}$

c) $(c^2 - 25) \div \dfrac{c-5}{c+5}$

b) $\dfrac{3y}{2(y+7)^2} \cdot \dfrac{y+7}{6y}$

d) $\dfrac{x}{x^2-1} \div \dfrac{2x^3}{1-x}$

Solution.

a) 4 and 2 share a common factor 2 while z and z^2 share a common factor z

$$\frac{4}{z^2} \cdot \frac{z}{2} = \frac{\overset{2}{\cancel{4}}}{\underset{z}{\cancel{z^2}}} \cdot \frac{\overset{1}{\cancel{z}}}{\cancel{2}}$$

$$= \frac{2}{z} \cdot \frac{1}{1}$$

$$= \frac{2}{z}$$

b) $3y$ and $6y$ share a common factor $3y$ while $(y + 7)$ and $(y + 7)^2$ share a common factor $(y+7)$

$$\frac{3y}{2(y+7)^2} \cdot \frac{y+7}{6y} = \frac{\overset{1}{\cancel{3y}}}{2\underset{(y+7)}{\cancel{(y+7)^2}}} \cdot \frac{\overset{1}{\cancel{(y+7)}}}{\underset{2}{\cancel{6y}}}$$

$$= \frac{1}{2(y+7)} \cdot \frac{1}{2}$$

$$= \frac{1}{4(y+7)}$$

Note. In this example, we could also have distributed the factor, 4, giving us $\frac{1}{4y+28}$. While both answers are acceptable, we recommend leaving expressions in their factored form.

c) Convert the division problem into a multiplication problem and factor $c^2 - 25$ first.

$$(c^2 - 25) \div \frac{c-5}{c+5} = \frac{c^2-25}{1} \cdot \frac{c+5}{c-5}$$

$$= \frac{(c-5)(c+5)}{1} \cdot \frac{c+5}{c-5}$$

$$= \frac{\overset{1}{\cancel{(c-5)}}(c+5)}{1} \cdot \frac{(c+5)}{\underset{1}{\cancel{(c-5)}}}$$

$$= \frac{(c+5)}{1} \cdot \frac{(c+5)}{1}$$

$$= (c+5)^2$$

Note. We recommend leaving the answer in factored form as $(c+5)^2$. If $(c+5)^2$ is to be expanded, recall that $(c+5)^2 = (c+5)(c+5) = c^2 + 10c + 25$.

d) Convert the division problem into a multiplication problem and factor $x^2 - 2x + 1$ first.

$$\frac{x}{x^2-2x+1} \div \frac{2x^3}{1-x} = \frac{x}{x^2-2x+1} \cdot \frac{1-x}{2x^3}$$

$$= \frac{x}{(x-1)(x-1)} \cdot \frac{-(x-1)}{2x^3}$$

$$= \frac{\overset{1}{\cancel{x}}}{\underset{1}{\cancel{(x-1)}}(x-1)} \cdot \frac{-(\overset{1}{\cancel{x-1}})}{\underset{x^2}{\cancel{2x^3}}}$$

$$= \frac{1}{(x-1)} \cdot \frac{-1}{2x^2}$$

$$= \frac{-1}{2x^2(x-1)} \quad \text{or} \quad -\frac{1}{2x^2(x-1)}$$

Problem 5 : | *Media/Class Example*

Perform the indicated operation and provide your final answer in simplified form.

a) $\dfrac{20}{n} \cdot \dfrac{3n^2}{5}$

d) $\dfrac{d^2-9}{d+3} \div (d-3)$

b) $(x^2+6x) \cdot \dfrac{x}{6+x}$

e) $\dfrac{y-4}{y} \div \dfrac{4-y}{2y}$

c) $\dfrac{n-7}{n^2-2n-35} \cdot \dfrac{9n+54}{10n+50}$

f) $\dfrac{7+k}{k^2-k-12} \div \dfrac{k^2+7k}{k^2-4k}$

Problem 6 : | *You Try*

Perform the indicated operation and provide your final answer in simplified form.

a) $\dfrac{9}{4m^2} \cdot \dfrac{m}{3}$

d) $y^2 \div \dfrac{y+4}{y}$

b) $\dfrac{7c}{5+c} \cdot (c^2 - 25)$

e) $\dfrac{6}{3a-1} \div \dfrac{24}{(3a-1)^2}$

c) $\dfrac{5-x}{x^2+3x} \cdot \dfrac{7x+21}{x^2-4x-5}$

f) $\dfrac{2n^2-12n+18}{n+7} \div (2n+6)$

Practice Problems: *Multiplying and Dividing Rational Expressions*

Perform the indicated operation and provide your final answer in simplified form.

1. $\dfrac{6}{7} \cdot \dfrac{5}{48}$

2. $\dfrac{13}{4} \cdot \dfrac{-3}{5}$

3. $\dfrac{55}{14} \cdot \dfrac{28}{15}$

4. $-\dfrac{9}{8} \cdot (-16)$

5. $\dfrac{4}{3} \div \left(-\dfrac{2}{9}\right)$

6. $\dfrac{\frac{14}{5}}{\frac{28}{5}}$

7. $-9 \div \dfrac{-25}{9}$

8. $\dfrac{\frac{36}{7}}{-4}$

Optional Challenge Questions. Perform the indicated operation and provide your final answer in simplified form.

9. $\dfrac{1}{2} \cdot \dfrac{2}{3} \cdot \dfrac{3}{4} \cdot \dfrac{4}{5} \cdot \dfrac{5}{6} \cdot \dfrac{6}{7} \cdot \dfrac{7}{8}$

10. $2 \cdot \dfrac{5}{4} \cdot \dfrac{11}{10} \cdot \dfrac{23}{22} \cdot \dfrac{47}{46}$

Perform the indicated operation and provide your final answer in simplified form.

11. $\dfrac{3}{x^2} \cdot \dfrac{5x}{6}$

12. $\dfrac{2y}{y-1} \cdot \dfrac{5}{4y}$

13. $(2z+6) \div (z^2+3z)$

14. $\dfrac{(n+2)^3}{5x} \cdot \dfrac{10}{n+2}$

15. $\dfrac{4w^2}{4+w} \cdot \dfrac{w^2-16}{w^3}$

16. $\dfrac{a^2+6a+9}{3} \div (3+a)$

17. $\dfrac{y-4}{3y^2} \cdot \dfrac{6y}{4-y}$

18. $\dfrac{-7}{c+5} \div \dfrac{3c}{2c-9}$

19. $\dfrac{6h}{7h+3} \div \dfrac{3}{(7h+3)^2}$

20. $\dfrac{2}{8+c} \cdot \dfrac{c^2-64}{2c-16}$

21. $\dfrac{v-1}{8} \cdot \dfrac{4}{v^2-11v+10}$

22. $\dfrac{g-5}{7-2g} \div \dfrac{5-g}{2g-7}$

23. $\dfrac{p-8}{p^2-12p+32} \div \dfrac{5}{p-4}$

24. $\dfrac{x^2-7x+10}{2-x} \cdot \dfrac{x+4}{x^2-x-20}$

25. $\dfrac{k}{k^2-k-12} \div \dfrac{k^2+4k}{k^2-16}$

Rescue Roody!

Help Roody find the errors and understand what went wrong in the following solutions.

26. Peform the indicated operation and simplify your answer: $\dfrac{x+3}{5} \cdot \dfrac{10}{3+x}$

   Solution.

$$\frac{x+3}{5} \cdot \frac{10}{3+x} = \frac{-1(3+x)}{\underset{1}{5}} \cdot \frac{10}{\underset{2}{(3+x)}}$$

$$= \frac{-1(\cancel{3}+x)}{\underset{1}{\cancel{5}}} \cdot \frac{\cancel{10}}{\underset{1}{(3\cancel{+}x)}}$$

$$= \frac{-1}{1} \cdot \frac{2}{1}$$

$$= -2$$

27. Perform the indicated operation and simplify your answer: $\dfrac{y+8}{y} \cdot \dfrac{y^2}{16}$

   Solution.

$$\frac{y+8}{y} \cdot \frac{y^2}{16} = \frac{\cancel{y}+8}{\cancel{y}} \cdot \frac{y^2}{16}$$

$$= \frac{\cancel{8}}{1} \cdot \frac{y^2}{\cancel{16}}$$

$$= \frac{1}{1} \cdot \frac{y^2}{2}$$

$$= \frac{y^2}{2}$$

10.3. ADDING AND SUBTRACTING RATIONAL EXPRESSIONS

Objective: To add and subtract rational expressions with and without common denominators.

Like Denominators

The steps for adding or subtracting rational expressions are the same as when working with fractions. When denominators are the same, we either add or subtract the numerators.

| **Problem 1 :** | *Worked Example* |

Perform the inidicated operation and simplify your final answer.

a) $\dfrac{5}{8} + \dfrac{7}{8}$
    b) $\dfrac{5}{2x} + \dfrac{7}{2x}$

Solution.

a) Since the denominators are the same, we add the numerators.

$$\frac{5}{8} + \frac{7}{8} = \frac{5+7}{8}$$
$$= \frac{12}{8}$$

Since 12 and 8 have a common factor of 4, our answer, $\dfrac{12}{8}$ can be simplified as $\dfrac{3}{2}$. Our final answer is $\dfrac{3}{2}$.

b) Since the denominators are the same, we add the numerators.

$$\frac{5}{2x} + \frac{7}{2x} = \frac{5+7}{2x}$$
$$= \frac{12}{2x}$$

Since 12 and 2 have a common factor of 2, our answer, $\dfrac{12}{2x}$ can be simplified as $\dfrac{6}{x}$. Our final answer is $\dfrac{6}{x}$.

Problem 2 : | *Media/Class Example*

Perform the indicated operation and simplify your final answer.

a) $\dfrac{14}{3y^2} + \dfrac{10}{3y^2}$

c) $\dfrac{2}{m^2-4} + \dfrac{m}{m^2-4}$

b) $\dfrac{3}{x+2} - \dfrac{7}{x+2}$

d) $\dfrac{4}{a-2} - \dfrac{2a}{a-2}$

Problem 3 : | *Worked Example*

Perform the indicated operation and simply your final answer: $\dfrac{2x+5}{x+3} - \dfrac{x-1}{x+3}$

   Solution.

Since the denominators are the same, we subtract numerators. Note that the minus sign in front of the second fraction changes the sign of every term in the numerator.

$$\frac{2x+5}{x+3} - \frac{x-1}{x+3} = \frac{(2x+5) - (x-1)}{x+3}$$
$$= \frac{2x+5-x+1}{x+3}$$
$$= \frac{x+6}{x+3}$$

Our final answer is $\dfrac{x+6}{x+3}$. This is already in simplified form.

Problem 4 : | *Media/Class Example*

Perform the indicated operation and simply your final answer.

a) $\dfrac{1}{y^2-16}+\dfrac{y-5}{y^2-16}$

b) $\dfrac{x+5}{x-6}-\dfrac{2x-1}{x-6}$

Problem 5 : | *Media/Class Example*

Perform the indicated operation and simply your final answer.

a) $\dfrac{x}{x+5}-\dfrac{5}{x+5}$

d) $\dfrac{3y}{y+1}-\dfrac{y+2}{y+1}$

b) $\dfrac{y}{6y+3}+\dfrac{y+1}{6y+3}$

e) $\dfrac{v^2}{v-5}-\dfrac{10v-25}{v-5}$

c) $\dfrac{b}{2a-2b}-\dfrac{a}{2a-2b}$

f) $\dfrac{8}{n+4}-\dfrac{4n}{n+4}+\dfrac{6n}{n+4}$

Unlike Denominators

What about when the denominators are not the same? We have to first build equivalent fractions with a common denominator.

Problem 6 : | _Worked Example_

Perform the indicated operation and simplify your final answer: $\dfrac{3}{10} - \dfrac{5}{12}$

Solution.

Since the denominators are different, we must create equivalent fractions having a common denominators for the original fractions. To do this, we find the Least Common Denominator (or LCD). We start with the prime factorization of the denominators, 10 and 12.

$$10 = 2 \cdot 5$$
$$12 = 2 \cdot 2 \cdot 3$$

To find the LCD, take all the factors from one denominator, then multiply any missing factors from the other denominator.

In this example, take the factors of 10 (which are $2 \cdot 5$) and multiply the missing factors from 12, which are $2 \cdot 3$. Therefore, the LCD is $2 \cdot 5 \cdot 2 \cdot 3 = 60$.

To create an equivalent fraction for $\dfrac{3}{10}$ with a denominator of 60, we multiply both numerator and denominator by the missing factors, $2 \cdot 3 = 6$:

$$\frac{3}{10} \cdot \frac{6}{6} = \frac{18}{60}$$

To create an equivalent fraction for $\dfrac{5}{12}$ with a denominator of 60, we multiply both numerator and denominator by the missing factor, 5:

$$\frac{5}{12} \cdot \frac{5}{5} = \frac{25}{60}$$

Consequently, $\dfrac{3}{10} - \dfrac{5}{12}$ becomes $\dfrac{18}{60} - \dfrac{25}{60}$. We now have common denominators, so we can subtract numerators.

$$\frac{3}{10} - \frac{5}{12} = \frac{18}{60} - \frac{25}{60}$$
$$= \frac{18-25}{60}$$
$$= \frac{-7}{60} \quad \text{or} \quad -\frac{7}{60}$$

Our final answer is $-\dfrac{7}{60}$. This cannot be simplified further.

Note. Why do we multiply both the numerator and denominator by the missing factors? We want to build equivalent fractions, meaning the new fraction has the same value as the original. Multiplying by 1, such as $\frac{6}{6}$ or $\frac{5}{5}$ does not change the value of the fraction.

We follow that same process with rational expressions.

| **Problem 7 :** | *Worked Example* |

Perform the indicated operation and simplify your final answer.

a) $\dfrac{4}{x^2}+\dfrac{6}{5x}$

b) $\dfrac{5}{y+2}-\dfrac{3}{4y}$

c) $\dfrac{5}{v^2-v}+\dfrac{6}{v-1}$

d) $\dfrac{7}{z-3}+\dfrac{2}{3-z}$

Solution.

a) The denominators are different. Let us do prime factorization for each.

$$x^2 = x\cdot x$$
$$5x = 5\cdot x$$

In this example, take the factors of x^2 (which are $x\cdot x$) and multiply the missing factor from $5x$, which is 5. This gives us the LCD: $x\cdot x\cdot 5=5x^2$.
Each denominator needs to be built up to the LCD, $5x^2$.

To create an equivalent fraction for $\dfrac{4}{x^2}$ with a denominator of $5x^2$, we multiply both numerator and denominator by the missing factor, 5.

$$\frac{4}{x^2}\cdot\frac{5}{5} = \frac{20}{5x^2}$$

To create an equivalent fraction for $\dfrac{6}{5x}$ with a denominator of $5x^2$, we multiply both numerator and denominator by the missing factor, x.

$$\frac{6}{5x}\cdot\frac{x}{x} = \frac{6x}{5x^2}$$

Consequently, $\dfrac{4}{x^2}+\dfrac{6}{5x}$ becomes $\dfrac{20}{5x^2}+\dfrac{6x}{5x^2}$. We now have common denominators, so we can add numerators.

$$\frac{4}{x^2}+\frac{6}{5x} = \frac{20}{5x^2}+\frac{6x}{5x^2}$$
$$= \frac{20+6x}{5x^2}$$
$$= \frac{6x+20}{5x^2}$$

Our final answer is $\dfrac{6x+20}{5x^2}$. We can factor the numerator $\dfrac{2(3x+10)}{5x^2}$ and determine that it cannot be simplified further. Either answer is acceptable.

b) The denominators are different. Let us do prime factorization of each.

$$y+2 = (y+2)$$
$$4y = 4 \cdot y$$

Note. $(y+2)$ is a prime binomial. It cannot be factored further. y and 2 are *terms* of the binomial, NOT factors.

In this example, take the factor of $(y+2)$ and multiply the missing factor from $4y$, which is $4y$. This gives us the LCD: $(y+2) \cdot 4 \cdot y = 4y(y+2)$.
Each denominator needs to be built up to the LCD, $4y(y+2)$.

To create an equivalent fraction for $\dfrac{5}{y+2}$ with a denominator of $4y(y+2)$, we multiply both numerator and denominator by the missing factor, $4y$.

$$\frac{5}{y+2} \cdot \frac{4y}{4y} = \frac{20y}{4y(y+2)}$$

To create an equivalent fraction for $\dfrac{3}{4y}$ with a denominator of $4y(y+2)$, we multiply numerator and denominator by the missing factor, $(y+2)$.

$$\frac{3}{4y} \cdot \frac{(y+2)}{(y+2)} = \frac{3(y+2)}{4y(y+2)}$$

Consequently, $\dfrac{5}{y+2} - \dfrac{3}{4y}$ becomes $\dfrac{20y}{4y(y+2)} - \dfrac{3(y+2)}{4y(y+2)}$. We now have common denominators, so we can subtract numerators.

$$\frac{5}{y+2} - \frac{3}{4y} = \frac{20y}{4y(y+2)} - \frac{3(y+2)}{4y(y+2)}$$
$$= \frac{20y - 3(y+2)}{4y(y+2)}$$
$$= \frac{20y - 3y - 6}{4y(y+2)}$$
$$= \frac{17y - 6}{4y(y+2)}$$

Our final answer is $\dfrac{17y-6}{4y(y+2)}$. This cannot be simplified further.

c) The denominators are different. Let us do prime factorization of each.

$$v^2 - v = v \cdot (v-1)$$
$$v-1 = (v-1)$$

Note. $(v-1)$ is a prime binomial. It cannot be factored further.

In this example, take the factors of $v \cdot (v-1)$ and multiply the missing factor from $(v-1)$, but there is none. Therefore, the LCD is $v \cdot (v-1) = v(v-1)$.

Each denominator needs to be built up to the LCD, $v(v-1)$.

The denominator of the rational expression $\dfrac{5}{v^2-v} = \dfrac{5}{v(v-1)}$ already has the LCD. We do not need to multiply numerator and denominator by the any missing factor.

To create an equivalent fraction for $\dfrac{6}{v-1}$ with a denominator of $v(v-1)$, we multiply numerator and denominator by the missing factor, v.

$$\frac{6}{v-1} \cdot \frac{v}{v} = \frac{6v}{v(v-1)}$$

Consequently, $\dfrac{5}{v^2-v} + \dfrac{6}{v-1}$ becomes $\dfrac{5}{v(v-1)} + \dfrac{6v}{v(v-1)}$. We now have common denominators, so we can add numerators.

$$
\begin{aligned}
\frac{5}{v^2-v} + \frac{6}{v-1} &= \frac{5}{v(v-1)} + \frac{6v}{v(v-1)} \\
&= \frac{5+6v}{v(v-1)} \\
&= \frac{6v+5}{v(v-1)}
\end{aligned}
$$

Our final answer is $\dfrac{6v+5}{v(v-1)}$. This cannot be simplified further.

d) The denominators are different. However, $3-z = -(z-3)$. The second rational expression becomes:

$$
\begin{aligned}
\frac{3}{3-z} &= \frac{3}{-(z-3)} \\
&= -\frac{3}{(z-3)}
\end{aligned}
$$

Consequently, $\dfrac{7}{z-3} + \dfrac{2}{3-z}$ becomes $\dfrac{7}{z-3} - \dfrac{2}{z-3}$. We now have common denominators, so we can add numerators.

$$
\begin{aligned}
\frac{7}{z-3} + \frac{2}{3-z} &= \frac{7}{z-3} - \frac{2}{z-3} \\
&= \frac{7-2}{z-3} \\
&= \frac{5}{z-3}
\end{aligned}
$$

Our final answer is $\dfrac{5}{z-3}$. This cannot be simplified further.

Problem 8 : | *Media/Class Example*

Perform the indicated operation and simplify your final answer.

a) $\dfrac{2}{15c}+\dfrac{5}{12c}$

d) $\dfrac{n}{n+3}-\dfrac{9}{n}$

b) $\dfrac{4}{3y}+\dfrac{7}{y^2}$

e) $\dfrac{3}{m^2+2m}+\dfrac{8}{m}$

c) $\dfrac{8}{k-2}-\dfrac{9}{2-k}$

f) $\dfrac{5}{2w}-3$

Problem 9 : | *You Try*

Perform the indicated operation and simplify your final answer.

a) $\dfrac{3}{c} + \dfrac{12}{5c}$

e) $\dfrac{1}{w+2} + \dfrac{1}{w-2}$

b) $\dfrac{5}{y^2} - \dfrac{3}{10y}$

f) $\dfrac{m}{m+5} - \dfrac{25}{m}$

c) $\dfrac{5}{x} + 2$

g) $\dfrac{8}{(p+3)^2} - \dfrac{2}{p+3}$

d) $\dfrac{3}{n} - \dfrac{4}{5n} + \dfrac{7}{2n}$

h) $\dfrac{3}{k} + \dfrac{3}{k^2 + 6k}$

Practice Problems: *Adding and Subtracting Rational Expressions*

Perform the indicated operation and simplify your answer.

1. $\dfrac{2}{a+3}+\dfrac{4}{a+3}$

2. $\dfrac{11}{y+2}-\dfrac{6}{y+2}$

3. $\dfrac{7}{3x}+\dfrac{5}{3x}$

4. $\dfrac{3m}{m-5}-\dfrac{15}{m-5}$

5. $\dfrac{6}{r-6}-\dfrac{r}{r-6}$

6. $\dfrac{t^2+4t}{t-1}+\dfrac{2t-7}{t-1}$

7. $\dfrac{x^2}{x-2}-\dfrac{6x-8}{x-2}$

8. $\dfrac{3}{4y-12}-\dfrac{y}{4y-12}$

9. $\dfrac{2w}{5}+\dfrac{7w}{4}$

10. $\dfrac{3}{y}-\dfrac{9}{2y}$

11. $\dfrac{3}{x}+\dfrac{4}{x^2}$

12. $\dfrac{6}{p}-\dfrac{14}{p^2}$

13. $\dfrac{x+5}{8}+\dfrac{x-3}{12}$

14. $\dfrac{a+2}{2}-\dfrac{a-4}{4}$

15. $\dfrac{2}{r}+\dfrac{3}{r-6}$

16. $\dfrac{7}{y+2}-\dfrac{5}{y}$

17. $\dfrac{8}{x-5}+\dfrac{3}{5-x}$

18. $\dfrac{m}{2-m}-\dfrac{4m}{m-2}$

19. $\dfrac{10}{w+2}-3$

20. $5+\dfrac{4}{k^2}$

21. $\dfrac{y}{(y+3)^2}+\dfrac{2}{y+3}$

22. $\dfrac{4}{a+1}-\dfrac{4}{(a+1)^2}$

23. $\dfrac{4}{n^2+5n}-\dfrac{1}{n}$

24. $\dfrac{2}{w}+\dfrac{6}{w^2-3w}$

Rescue Roody!

What did Roody do wrong in each of these problems? Identify the error(s) then perform the operation correctly.

25. Perform the indicated operation and simplify the answer: $\dfrac{1}{a}+\dfrac{1}{5}$

Solution.

$$\frac{1}{a}+\frac{1}{5}=\frac{2}{a+5}$$

26. Perform the indicated operation and simplify the answer: $\dfrac{1}{x} + \dfrac{1}{x+3}$

Solution.

The LCD is $x+3$

$$\frac{1}{x} + \frac{1}{x+3} = \frac{3}{x+3} + \frac{1}{x+3}$$
$$= \frac{4}{x+3}$$

27. Perform the indicated operation and simplify the answer: $\dfrac{x^2}{x-2} - \dfrac{6x-8}{x-2}$

Solution.

$$\frac{x^2}{x-2} - \frac{6x-8}{x-2} = \frac{x^2-6x-8}{x-2}$$
$$= \frac{(x-4)(x-2)}{(x-2)}$$
$$= x-4$$

10.4. COMPLEX FRACTIONS

Objective: To simplify complex fractions

An expression of the form $\dfrac{N}{D}$ is called a **complex fraction**, when the numerator N, or the denominator, D, or both, contain a fraction. Examples of complex fractions are

$$\dfrac{\dfrac{4}{3}}{9}, \qquad \dfrac{\dfrac{3}{4}-5}{\dfrac{3}{7}}, \qquad \dfrac{\dfrac{2}{3}}{\dfrac{1}{4}}, \qquad \dfrac{1+\dfrac{1}{x}}{x}$$

Our goal is to simplify complex fractions. Recall that if we have a fraction where the

denominator is a fraction itself, as in $\dfrac{\dfrac{5}{7}}{\dfrac{3}{2}}$, this is equivalent to multipling the numerator

by the reciprocal of the denominator. In our example this becomes

$$\dfrac{\dfrac{5}{7}}{\dfrac{3}{2}} = \dfrac{5}{7}\cdot\dfrac{2}{3}$$

$$= \dfrac{10}{21}$$

Problem 1 :	*Worked Example*

Simplify the following expressions

a) $\dfrac{\dfrac{2}{3}+1}{\dfrac{1}{2}}$
b) $\dfrac{\dfrac{2}{5}}{4}$
c) $\dfrac{3}{\dfrac{5}{4}+2}$
d) $\dfrac{\dfrac{5}{9}+\dfrac{2}{3}}{\dfrac{1}{3}}$

Solution.

a) Method 1

We can combine the terms in the numerator and denominator first.

$$\frac{\frac{2}{3}+1}{\frac{1}{2}} = \frac{\frac{2}{3}+\frac{3}{3}}{\frac{1}{2}}$$

$$= \frac{\frac{5}{3}}{\frac{1}{2}} \qquad \text{multiply by the reciprocal of the denominator}$$

$$= \frac{5}{3}\cdot\frac{2}{1} \qquad \text{multiply}$$

$$= \frac{10}{3} \qquad \text{our solution}$$

a) Method 2

We can multiply the numerator and denominator by the LCD of all fractions involved. In this case, the LCD of the fractions $\frac{2}{3}$ and $\frac{1}{2}$ is 6. Multiplying by $\frac{6}{6}$ is equivalent to multiplying by 1.

$$\frac{\frac{2}{3}+1}{\frac{1}{2}} = \frac{\left(\frac{2}{3}+1\right)}{\left(\frac{1}{2}\right)}\cdot\frac{6}{6} \qquad \text{multiply numerator and denominator by 6}$$

$$= \frac{\left(\frac{2}{3}\right)6+(1)6}{\left(\frac{1}{2}\right)6} \qquad \text{simplify}$$

$$= \frac{4+6}{3}$$

$$= \frac{10}{3} \qquad \text{our solution}$$

We illustrated two different methods to simplify complex fractions. Method 1 combines the fractions, while Method 2 clears the fractions. With practice you will develop the skill to decide with method is more efficient in each case. From now on we will choose one or the other method to simplify.

b) We will use method 1, multiply the numerator and denominator by 5:

$$\frac{\frac{2}{5}}{4} = \frac{\frac{2}{5}}{4}\frac{5}{5}$$

$$= \frac{\frac{2}{5}(5)}{4(5)} \qquad \text{Simplify}$$

$$= \frac{2}{20} \qquad \text{Simplify}$$

$$= \frac{1}{10} \qquad \text{Our solution}$$

c) We will use Method 1:

$$\frac{3}{\frac{5}{4}+2} = \frac{3}{\frac{5}{4}+\frac{8}{4}} \qquad \text{Combine the terms in the denominator}$$

$$= \frac{3}{\frac{13}{4}} \qquad \text{Multiply by the reciprocal}$$

$$= \frac{3}{1} \cdot \frac{4}{13} \qquad \text{Multiply}$$

$$= \frac{12}{13} \qquad \text{Our solution.}$$

d) We will use method 2:

$$\frac{\frac{5}{9}+\frac{2}{3}}{\frac{1}{3}} = \frac{\frac{5}{9}+\frac{2}{3}}{\frac{1}{3}} \cdot \frac{9}{9} \qquad \text{multiply by the LCD (9)}$$

$$= \frac{\left(\frac{5}{9}\right)9+\left(\frac{2}{3}\right)9}{\left(\frac{1}{3}\right)9} \qquad \text{simplify}$$

$$= \frac{5+6}{3} \qquad \text{add}$$

$$= \frac{11}{3} \qquad \text{our solution.}$$

Problem 2 : | *Media/Class Example*

a) $\dfrac{\dfrac{2}{7}}{\dfrac{4}{5}}$

b) $\dfrac{\dfrac{5}{9}}{8}$

c) $\dfrac{3-\dfrac{1}{2}}{\dfrac{3}{5}}$

d) $\dfrac{\dfrac{3}{7}}{\dfrac{1}{6}+\dfrac{3}{4}}$

Problem 3 : | *You Try*

Simplify the following fractions

a) $\dfrac{\dfrac{2}{5}}{\dfrac{3}{2}}$

b) $\dfrac{2}{\dfrac{1}{4}+2}$

c) $\dfrac{\dfrac{3}{4}-\dfrac{2}{3}}{\dfrac{1}{6}}$

We will now focus on simplifying complex fractions involving variables. The same principles apply as before, and we can chose either of the two methods shown above to simplify a given expression.

| **Problem 4 :** | *Worked Example* |

Simplify the expressions

a) $\dfrac{\dfrac{4}{x}}{\dfrac{6}{x^2}}$

b) $\dfrac{x+\dfrac{1}{x}}{2}$

c) $\dfrac{3}{1-\dfrac{5}{x}}$

Solution.

a) We will multiply by the reciprocal of the denominator (Method 1):

$$\dfrac{\dfrac{4}{x}}{\dfrac{6}{x^2}} = \dfrac{4}{x} \cdot \dfrac{x^2}{6}$$

$$= \dfrac{4x}{6} \qquad \text{simplify}$$

$$= \dfrac{2x}{3} \qquad \text{our solution}$$

b) We will use method 2

$$\dfrac{x+\dfrac{1}{x}}{2} = \dfrac{\left(x+\dfrac{1}{x}\right)x}{2} \qquad \text{multiply by the LCD: } x$$

$$= \dfrac{x \cdot x + \left(\dfrac{1}{x}\right)x}{2x} \qquad \text{write the denominator as a fraction}$$

$$= \dfrac{x^2+1}{2x} \qquad \text{our solution}$$

c) We will use method **1**

$$\frac{3}{1-\dfrac{5}{x}} = \frac{3}{\dfrac{x}{x}-\dfrac{5}{x}}$$

$$= \frac{\dfrac{3}{1}}{\dfrac{x-5}{x}}$$

$$= \frac{3}{1} \cdot \frac{x}{x-5}$$

$$= \frac{3x}{x-5} \qquad \text{our solution}$$

Problem 5 : | *Class/Media Example*

Simplify the following expressions:

a) $\dfrac{\dfrac{4}{3y}}{\dfrac{2}{5y^2}}$

b) $\dfrac{1+\dfrac{3}{x}}{x-\dfrac{1}{4}}$

c) $\dfrac{\dfrac{3}{2x}}{1+\dfrac{1}{x}}$

d) $\dfrac{x-1}{x-\dfrac{1}{4}}$

Problem 6 : | *You Try*

Simplify the following expressions:

a) $\dfrac{8x}{\dfrac{4}{x}}$

b) $\dfrac{1+\dfrac{1}{x}}{1-\dfrac{1}{x}}$

c) $\dfrac{\dfrac{3}{2}}{1+\dfrac{1}{3x}}$

d) $\dfrac{1+\dfrac{5}{x}}{x+2}$

| Problem 7 : | Worked Example |

Simplify the following expressions:

$$\text{a) } \dfrac{1+\dfrac{1}{x}}{x-\dfrac{1}{x}} \qquad\qquad \text{b) } \dfrac{3-\dfrac{3}{x}}{1-\dfrac{1}{x^2}}$$

Solution.

We will use method 2:

$$\dfrac{1+\dfrac{1}{x}}{x-\dfrac{1}{x}} = \dfrac{1+\dfrac{1}{x}}{x-\dfrac{1}{x}} \cdot \dfrac{x}{x} \qquad \text{Multiply by the LCD, } x$$

$$= \dfrac{x+1}{x^2-1} \qquad \text{simplify}$$

$$= \dfrac{(x+1)}{(x+1)(x-1)}$$

$$= \dfrac{1}{x-1} \qquad \text{our solution}$$

b) We will use method 2.

$$\dfrac{3-\dfrac{3}{x}}{1-\dfrac{1}{x^2}} = \dfrac{3-\dfrac{3}{x}}{1-\dfrac{1}{x^2}} \cdot \dfrac{x^2}{x^2} \qquad \text{Multiply by the LCD, } x^2$$

$$= \dfrac{3x^2-3x}{x^2-1} \qquad \text{Simplify}$$

$$= \dfrac{3x(x-1)}{(x+1)(x-1)}$$

$$= \dfrac{3x}{(x+1)} \qquad \text{our solution}$$

Problem 8 : | Class/Media Example

Simplify the following expressions:

a) $\dfrac{1+\dfrac{1}{x}}{x-\dfrac{1}{4}}$

b) $\dfrac{x-\dfrac{9}{x}}{1+\dfrac{3}{x}}$

Problem 9 : | You Try

Simplify the following expressions:

a) $\dfrac{x-\dfrac{4}{x}}{1-\dfrac{2}{x}}$

b) $\dfrac{x}{2x+\dfrac{3}{x}}$

c) $\dfrac{3-\dfrac{2}{x^2}}{2+\dfrac{3}{x}}$

Practice Problems: *Complex Fractions*

1. $\dfrac{\dfrac{1}{2x}}{\dfrac{3}{4x^2}}$

2. $\dfrac{\dfrac{x}{5}}{\dfrac{2}{x^2}}$

3. $\dfrac{\dfrac{3}{7x}}{\dfrac{6x}{5}}$

4. $\dfrac{\dfrac{4x}{9}}{8x}$

5. $\dfrac{\dfrac{2}{x}+1}{\dfrac{5}{6}}$

6. $\dfrac{1-\dfrac{x}{8}}{4}$

7. $\dfrac{4x-\dfrac{2}{5}}{\dfrac{2}{3}}$

8. $\dfrac{3x}{\dfrac{6x}{5}-1}$

9. $\dfrac{2}{\dfrac{1}{x}-\dfrac{1}{9}}$

10. $\dfrac{\dfrac{x}{2}}{x+\dfrac{3}{5}}$

11. $\dfrac{\dfrac{7x}{2}}{5-\dfrac{3}{2x}}$

12. $\dfrac{1+\dfrac{2}{x}}{\dfrac{2}{3x}+1}$

13. $\dfrac{2-\dfrac{2}{x}}{3+\dfrac{1}{x^2}}$

14. $\dfrac{1}{1-\dfrac{1}{x}}$

15. $\dfrac{x+6}{\dfrac{1}{3}+\dfrac{1}{x}}$

16. $\dfrac{\dfrac{2}{x}}{\dfrac{1}{x}-5x}$

17. $\dfrac{\dfrac{3}{7}-x}{\dfrac{3}{x}+2}$

18. $\dfrac{\dfrac{y}{9}-\dfrac{3}{y}}{\dfrac{9}{y}}$

19. $\dfrac{\dfrac{1}{a}+\dfrac{1}{8}}{\dfrac{1}{a}-\dfrac{1}{8}}$

20. $\dfrac{1+\dfrac{2}{c}}{1-\dfrac{4}{c^2}}$

10.5. SOLVING RATIONAL EQUATIONS

Objective: Solve rational equations and proportions.

When solving rational equations, we will clear the fraction by multiplying each fraction by the lowest common denominator (or LCD). This is basically the same strategy as solving linear equations with fractions. However, with solving rational equations, since some expression may contain a variable in the denominator, we have to be careful that our solution does not make any rational expression undefined.

The next two examples will show you the difference and similarity between how a linear equation and a rational equation look and how each is solved is.

Problem 1 : | *Worked Example*

Solve $\dfrac{2}{3}x + \dfrac{7}{6} = \dfrac{3}{4}x$, for x

> **Solution.**

This is a linear equation. Note that the position of the variable, x. Each fraction will never be undefined. Therefore, there are no values that have to be excluded. LCD for the equation is 12.

$$\frac{2}{3}x + \frac{7}{6} = \frac{3}{4}x \qquad \text{Multiply each side by the LCD, 12}$$

$$12\left(\frac{2}{3}x + \frac{7}{6}\right) = 12\left(\frac{3}{4}x\right) \qquad \text{Distribute}$$

$$12\left(\frac{2}{3}x\right) + 12\left(\frac{7}{6}\right) = 12\left(\frac{3}{4}x\right) \qquad \text{Multiply and simplify}$$

$$\overset{4}{\cancel{12}}\left(\frac{2}{\cancel{3}_1}x\right) + \overset{2}{\cancel{12}}\left(\frac{7}{\cancel{6}_1}\right) = \overset{3}{\cancel{12}}\left(\frac{3}{\cancel{4}_1}x\right)$$

$$8x + 14 = 9x \qquad \text{Solve the linear equation}$$

$$14 = x$$

Verify that we have the correct solution.

$$\frac{2}{3}(14) + \frac{7}{6} \overset{?}{=} \frac{3}{4}(14)$$

$$\frac{28}{3} + \frac{7}{6} \overset{?}{=} \frac{21}{2}$$

$$\frac{56}{6} + \frac{7}{6} \overset{?}{=} \frac{21}{2}$$

$$\frac{63}{6} = \frac{21}{2} \quad \checkmark$$

Therefore our solution is $x = 14$.

Problem 2 : | *Worked Example*

Solve $\dfrac{2}{3x}+\dfrac{7}{6}=\dfrac{3}{x}$ for x

Solution.

This is a rational equation. Note the position of the variable, x. LCD for the equation is $6x$. When the value of x is 0, the rational expressions $\dfrac{2}{3x}$ and $\dfrac{3}{x}$ become undefined. Therefore, $x = 0$ has to be excluded from the solution.

$$\dfrac{2}{3x}+\dfrac{7}{6} = \dfrac{3}{x} \qquad \text{Multiply each side by the LCD, } 6x$$

$$6x\left(\dfrac{2}{3x}+\dfrac{7}{6}\right) = 6x\left(\dfrac{3}{x}\right) \qquad \text{Distribute}$$

$$6x\left(\dfrac{2}{3x}\right)+6x\left(\dfrac{7}{6}\right) = 6x\left(\dfrac{3}{x}\right) \qquad \text{Multiply and simplify}$$

$$\overset{2}{\cancel{6x}}\left(\dfrac{2}{\cancel{3x}}\right)+\overset{x}{\cancel{6x}}\left(\dfrac{7}{\cancel{6}}\right) = \overset{6}{\cancel{6x}}\left(\dfrac{3}{\cancel{x}}\right)$$

$$4+7x = 18 \qquad \text{Solve the linear equation}$$

$$7x = 14$$

$$x = 2$$

Verify that we have the correct solution.

$$\dfrac{2}{3(2)}+\dfrac{7}{6} \overset{?}{=} \dfrac{3}{2}$$

$$\dfrac{1}{3}+\dfrac{7}{6} \overset{?}{=} \dfrac{3}{2}$$

$$\dfrac{2}{6}+\dfrac{7}{6} \overset{?}{=} \dfrac{3}{2}$$

$$\dfrac{9}{6} = \dfrac{3}{2} \quad \checkmark$$

Therefore our solution is $x = 2$.

Problem 3 : | *Media/Class Example*

Identify if the equation is linear or rational then solve for x.

a) $\dfrac{x}{5} - \dfrac{7}{15} = \dfrac{2}{3}x$

b) $\dfrac{4}{x} + 3 = \dfrac{5}{2x}$

Problem 4 : | *You Try*

Identify if the equation is linear or rational then solve for x.

a) $\dfrac{13}{8}x - \dfrac{1}{4} = \dfrac{1}{16} + x$

b) $\dfrac{1}{x} - \dfrac{1}{2} = \dfrac{5}{4x}$

Strategy for Solving Rational Equations:

1. Find the lowest common denominator (LCD).

2. Identify the values which make each term undefined. These values have to be excluded from the solution.

3. Multiply each term by the LCD.

4. Solve the resulting equation.

5. Verify that the solution is correct.

6. State your solution(s).

Let us apply the strategy to solve the following rational equations.

Problem 5 : | *Worked Example*

Solve $\dfrac{5}{2p-1}=\dfrac{3}{p}$ for p

Solution.

LCD is $p(2p-1)$ such that $p \neq 0$ because whe $p=0$, the term $\dfrac{3}{p}$ becomes undefined and $p \neq \dfrac{1}{2}$ because when $p=\dfrac{1}{2}$, the term $\dfrac{5}{2p-1}$ becomes undefined.

$$\dfrac{5}{2p-1} = \dfrac{3}{p} \qquad \text{Multiply each side by the LCD, } p(2p-1)$$

$$p(2p-1)\left(\dfrac{5}{2p-1}\right) = p(2p-1)\left(\dfrac{3}{p}\right) \qquad \text{Multiply and simplify}$$

$$p(2\overset{1}{\cancel{p}}\cancel{-}1)\left(\dfrac{5}{2\cancel{p}\cancel{-}1}\right) = \overset{1}{\cancel{p}}(2p-1)\left(\dfrac{3}{\underset{1}{\cancel{p}}}\right)$$

$$5p = 3(2p-1) \qquad \text{Solve the linear equation}$$

$$5p = 6p-3$$

$$-p = -3$$

$$p = 3$$

Let's verify that solution is correct.

When $p=3$:

$$\frac{5}{2(3)-1} \overset{?}{=} \frac{3}{(3)}$$

$$\frac{5}{6-1} \overset{?}{=} 1$$

$$\frac{5}{5} = 1 \quad \checkmark$$

Therefore $p=3$ is our solution.

| Problem 6 : | *Worked Example* |

Solve $x+5=\dfrac{6}{x}$ for x

Solution.

LCD is x such that $x \neq 0$ because when $x=0$, the term, $\dfrac{6}{x}$ is undefined.

$$x+5 = \frac{6}{x} \qquad \text{Multiply each side by the LCD, } x$$

$$x(x+5) = x\left(\frac{6}{x}\right) \qquad \text{Multiply and simplify}$$

$$x(x)+x(5) = \overset{1}{\cancel{x}}\left(\frac{6}{\underset{1}{\cancel{x}}}\right)$$

$$x^2+5x = 6 \qquad \text{Solve the quadratic equation}$$

$$x^2+5x-6 = 0$$

$$(x+6)(x-1) = 0$$

$$x+6=0 \quad \text{or} \quad x-1=0$$

$$x=-6 \quad \text{or} \quad x=1$$

Verify that the solution is correct.

When $x=-6$:   When $x=1$:

$$(-6)+5 \overset{?}{=} \frac{6}{(-6)} \qquad\qquad (1)+5 \overset{?}{=} \frac{6}{(1)}$$

$$1 = 1 \quad \checkmark \qquad\qquad\qquad 6 = 6 \quad \checkmark$$

Both answers check. Therefore, our solution is $x=-6$ or $x=1$.

Problem 7 : | *Worked Example*

Solve $3x - \dfrac{10x}{x+2} = \dfrac{x^2}{x+2}$ for x

Solution.

LCD is $(x+2)$ such that $x \neq -2$ because when $x=-2$, the terms $\dfrac{10x}{x+2}$ and $\dfrac{x^2}{x+2}$ are undefined.

$$3x - \frac{2x}{x+2} = \frac{x^2}{x+2} \qquad \text{Multiply each side by the LCD, } (x+2)$$

$$(x+2)\left(3x - \frac{2x}{x+2}\right) = (x+2)\left(\frac{x^2}{x+2}\right) \qquad \text{Distribute}$$

$$(x+2)3x - (x+2)\frac{2x}{x+2} = (x+2)\left(\frac{x^2}{x+2}\right) \qquad \text{Multiply and simplify}$$

$$(x+2)\,3x - (x\!\!\!/+2)\frac{\overset{1}{2x}}{\underset{1}{x\!\!\!/+2}} = (x\!\!\!/+2)\frac{\overset{1}{x^2}}{\underset{1}{x\!\!\!/+2}}$$

$$3x(x+2) - 2x = x^2 \qquad \text{Solve the quadratic equation}$$

$$3x^2 + 6x - 2x = x^2$$

$$2x^2 + 4x = 0$$

$$2x(x+2) = 0$$

$$2x = 0 \qquad \text{or} \qquad x+2=0$$

$$x = 0 \qquad \text{or} \qquad x = -2$$

$x = -2$ is an extraneous solution since $x = -2$ makes the rational expressions, $\dfrac{10x}{x+2}$ and $\dfrac{x^2}{x+2}$ undefined. Verify that $x = 0$ is correct.

When $x = 0$:

$$3(0) - \frac{10\,(0)}{(0)+2} \overset{?}{=} \frac{(0)^2}{(0)+2}$$

$$0 - \frac{0}{2} \overset{?}{=} \frac{0}{2}$$

$$0 - 0 = 0 \qquad \checkmark$$

Therefore, we have only one solution, $x = 0$.

Problem 8 : | *Worked Example*

Solve $\dfrac{2y}{y+1} = \dfrac{3}{y-1}$ for y

Solution.

LCD is $(y+1)(y-1)$ such that $y \neq -1$ because when $y=-1$, the term $\dfrac{2y}{y+1}$ becomes undefined and $y \neq 1$ because when $y=1$, the term $\dfrac{3}{y-1}$ becomes undefined.

$$\dfrac{2y}{y+1} = \dfrac{3}{y-1} \qquad \text{Multiply each term by the LCD}$$

$$(y+1)(y-1)\dfrac{2y}{y+1} = (y+1)(y-1)\dfrac{3}{y-1} \qquad \text{Multiply and simplify}$$

$$\overset{1}{(\cancel{y+1})}(y-1)\dfrac{2y}{\underset{1}{\cancel{y+1}}} = (y+1)\overset{1}{(\cancel{y-1})}\dfrac{3}{\underset{1}{\cancel{y-1}}}$$

$$2y(y-1) = 3(y+1) \qquad \text{Solve the quadratic equation}$$

$$2y^2 - 2y = 3y+3$$

$$2y^2 - 5y - 3 = 0$$

$$(2y+1)(y-3) = 0$$

$$\begin{array}{ccc} 2y+1=0 & \text{or} & y-3=0 \\ y=-\dfrac{1}{2} & \text{or} & y=3 \end{array}$$

Verify that the solution is correct.

When $y=-\dfrac{1}{2}$:

$$\dfrac{2\left(-\frac{1}{2}\right)}{\left(-\frac{1}{2}\right)+1} \overset{?}{=} \dfrac{3}{\left(-\frac{1}{2}\right)-1}$$

$$\dfrac{-1}{\frac{1}{2}} \overset{?}{=} \dfrac{3}{-\frac{3}{2}}$$

$$-2 = -2 \quad \checkmark$$

When $y=3$:

$$\dfrac{2(3)}{(3)+1} \overset{?}{=} \dfrac{3}{(3)-1}$$

$$\dfrac{6}{4} = \dfrac{3}{2} \quad \checkmark$$

Both answers check. Therefore our solution is $y=-\dfrac{1}{2}$ or $y=3$.

Problem 9 : | *Media/Class Example*

Solve the following equations.

a) $\dfrac{8}{n+4} = \dfrac{6}{n-1}$

d) $\dfrac{x^2}{x-1} = 2x + \dfrac{1}{x-1}$

b) $3y = \dfrac{1}{y} - \dfrac{1}{2}$

e) $\dfrac{1}{p^2} + \dfrac{1}{p} = 2$

c) $\dfrac{2m}{m-3} = \dfrac{m}{m+1}$

f) $\dfrac{n}{n-3} - \dfrac{4}{5(n-3)} = \dfrac{1}{2}$

Problem 10 : | *You Try*

Solve the following equations.

a) $\dfrac{2}{p-3}=\dfrac{5}{p}$

d) $m+\dfrac{8}{m-4}=\dfrac{5m}{m-4}$

b) $\dfrac{x-3}{x}=5-\dfrac{3}{x}$

e) $3+\dfrac{1}{y}=\dfrac{2}{y^2}$

c) $\dfrac{2n-1}{n}=\dfrac{2n}{n+1}$

f) $\dfrac{5}{2(a+1)}-3=\dfrac{2a}{a+1}$

Practice Problems: *Solving Rational Equations*

Solve the following equations:

1. $\dfrac{5}{3} = \dfrac{1}{2} + \dfrac{7}{x}$

2. $\dfrac{4}{m} = \dfrac{8}{m-1}$

3. $\dfrac{6}{y+2} = 9$

4. $5 - \dfrac{11}{3z} = \dfrac{2}{3}$

5. $\dfrac{3}{n} = \dfrac{6}{2n-1}$

6. $\dfrac{y}{5(y-1)} + \dfrac{3}{5} = \dfrac{1}{y-1}$

7. $\dfrac{x-3}{2x} = 5 - \dfrac{3}{x}$

8. $\dfrac{4}{p-2} = \dfrac{2p}{p-2} - 1$

9. $\dfrac{c}{9} = \dfrac{4}{c}$

10. $\dfrac{1}{y} + 2y = 3$

11. $\dfrac{x}{3} = \dfrac{4}{x-1}$

12. $a - \dfrac{12}{a} = 4$

13. $\dfrac{n}{n+1} = \dfrac{1}{2n}$

14. $\dfrac{2w}{w-3} - w = \dfrac{6}{w-3}$

15. $\dfrac{p-4}{p-1} = \dfrac{12}{3-p}$

16. $x = \dfrac{10}{x+1} + 8$

17. $\dfrac{2}{3(a-1)} - 1 = \dfrac{4a}{a-1}$

18. $\dfrac{3}{c^2} + \dfrac{1}{c} = 2$

CHAPTER 10 ASSESSMENT

Evaluate the expression for the given value. Simplify your answers.

1. $\dfrac{w}{w-3}$, for $w=0$

2. $\dfrac{a^2-a}{a^2-9a+8}$, for $a=-1$

Find all values of the variable that make the given rational expression undefined.

3. $\dfrac{4}{k+3}$

4. $\dfrac{3g^2-6g}{12}$

5. $\dfrac{y}{y^2-9}$

6. $\dfrac{c^2-4}{c^2+6c-8}$

Simplify the given rational expression.

7. $\dfrac{15b^2}{20b}$

8. $\dfrac{12+4k}{5k+15}$

9. $\dfrac{5-w}{w^2-3w-10}$

10. $\dfrac{m^2+4m+4}{2m+m^2}$

Perform the indicated operation and provide your final answer in simplified form.

11. $\dfrac{2-6x}{x^2+4x} \cdot \dfrac{7x}{3x^2+5x-2}$

12. $\dfrac{y^2+y-6}{y^2-9} \div (4x-12)$

13. $\dfrac{x-6}{6} + \dfrac{x-2}{3}$

14. $\dfrac{9}{n+3} - \dfrac{n^2}{n+3}$

15. $\dfrac{2m-5}{m-2} + \dfrac{m-3}{2-m}$

16. $\dfrac{5}{y^2-3y} - \dfrac{2}{y}$

Simplify the complex fraction.

17. $\dfrac{3+\dfrac{1}{2}}{\dfrac{7}{4}}$

18. $\dfrac{\dfrac{5}{n}-1}{\dfrac{5}{n}}$

19. $\dfrac{6m}{\dfrac{3}{m}}$

20. $\dfrac{7+\dfrac{1}{x}}{1+7x}$

Solve the following equations.

21. $\dfrac{a+1}{a-1} = \dfrac{3}{2}$

22. $\dfrac{9}{x-3} = \dfrac{3x}{x-3} - 1$

23. $\dfrac{2}{y} = \dfrac{y-3}{2}$

24. $\dfrac{5}{p} - \dfrac{4}{p^2} = 1$

Appendix A

Review of Arithmetic

A.1. Integers

Objective: Add, Subtract, Multiply and Divide Positive and Negative Numbers.

The ability to work comfortably with signed numbers is essential to success in algebra. For this reason, we will do a quick review of adding, subtracting, multiplying and dividing integers. **Integers** are all the whole numbers, zero, and their opposites (negatives). As this is intended to be a review of integers, the descriptions and examples will not be as detailed as a normal lesson.

World View Note: The first set of rules for working with negative numbers was written out by the Indian mathematician Brahmagupta.

When adding integers, we have two cases to consider. The first is if the signs match, both positive or both negative. If the signs match we will add the numbers together and keep the sign. This is illustrated in the following examples.

Problem 1 :	*Worked Example*

Add $-5 + (-3)$

 Solution.

$$-5 + (-3) \quad \text{Same sign, find the sum}$$
$$= -8 \quad \text{Our Solution}$$

Problem 2 :	*Worked Example*

Add $-7 + (-5)$

 Solution.

$$-7 + (-5) \quad \text{Same sign, find the sum}$$
$$= -12 \quad \text{Our Solution}$$

If the signs don't match, one positive and one negative number, we will subtract the numbers (as if they were all positive) and then use the sign from the larger number. This means if the larger number is positive, the answer is positive. If the larger number is negative, the answer is negative. This is shown in the following examples.

Problem 3 : | *Worked Example*

Add $-7 + 2$

  Solution.

$$-7 + 2 \quad \text{Different signs, find the difference}$$
$$= -5 \quad \text{Our Solution}$$

Problem 4 : | *Worked Example*

Add $-4 + 6$

  Solution.

$$-4 + 6 \quad \text{Different signs, find the difference}$$
$$= 2 \quad \text{Our Solution}$$

Problem 5 : | *Worked Example*

Add $4 + (-3)$

  Solution.

$$4 + (-3) \quad \text{Different signs, find the difference}$$
$$= 1 \quad \text{Our Solution}$$

Problem 6 : | *Worked Example*

Add $7 + (-10)$

  Solution.

$$7 + (-10) \quad \text{Different signs, find the difference}$$
$$= -3 \quad \text{Our Solution}$$

For subtraction of negatives we will change the problem to an addition problem which we can then solve using the above methods. The way we change a subtraction to an addition is to add the opposite of the number after the subtraction sign. Often this method is refered to as "add the opposite." This is illustrated in the following examples.

Problem 7 : | *Worked Example*

Subtract $9 - (-4)$

  Solution.

$$9 - (-4) \quad \text{Add the opposite of } -4$$
$$= 9 + 4 \quad \text{Same signs, find the sum}$$
$$= 13 \quad \text{Our Solution}$$

Problem 8 : | *Worked Example*

Subtract $-6-(-2)$

> **Solution.**

$$
\begin{aligned}
-6-(-2) \quad & \text{Add the opposite of} -2 \\
=-6+2 \quad & \text{Different signs, find the difference} \\
=-4 \quad & \text{Our Solution}
\end{aligned}
$$

Multiplication and division of integers both work in a very similar pattern. The short description of the process is we multiply and divide like normal, if the signs match (both positive or both negative) the answer is positive. If the signs don't match (one positive and one negative) then the answer is negative. This is shown in the following examples.

Problem 9 : | *Worked Example*

Multiply $(4)(-6)$

> **Solution.**

$$
\begin{aligned}
(4)(-6) \quad & \text{Signs do not match, answer is negative} \\
=-24 \quad & \text{Our Solution}
\end{aligned}
$$

Problem 10 : | *Worked Example*

Divide $\dfrac{-36}{-9}$

> **Solution.**

$$
\dfrac{-36}{-9} \quad \text{Signs match, answer is positive}
$$

$$
=4 \quad \text{Our Solution}
$$

Problem 11 : | *Worked Example*

Multiply $-2(-6)$

> **Solution.**

$$
\begin{aligned}
-2(-6) \quad & \text{Signs match, answer is positive} \\
=12 \quad & \text{Our Solution}
\end{aligned}
$$

Problem 12 : | *Worked Example*

Divide $\dfrac{15}{-3}$

> **Solution.**

$$
\dfrac{15}{-3} \quad \text{Signs do not match, answer is negative}
$$

$$
=-5 \quad \text{Our Solution}
$$

A few things to be careful of when working with integers. First, be sure not to confuse a problem like $-3 - 8$ with $-3(-8)$. The second problem is a multiplication problem because there is nothing between -3 and the parenthesis. If there is no operation written in between the parts, then that means multiplication. However, $-3 - 8$ is a subtraction problem because the subtraction separates the -3 from what comes after it. Another item to watch out for is to be careful not to mix up the pattern for adding and subtracting integers with the pattern for multiplying and dividing integers. They can look very similar. For example, if the signs match on addition, we keep the negative, $-3 + (-7) = -10$, but if the signs match on multiplication, the answer is positive, $(-3)(-7) = 21$.

Practice: *Integers*

Evaluate each expression.

1) $1 - 3$

3) $(-6) - (-8)$

5) $(-3) - 3$

7) $3 - (-5)$

9) $(-7) - (-5)$

11) $3 - (-1)$

13) $6 - 3$

15) $(-5) + 3$

17) $2 - 3$

19) $(-8) - (-5)$

21) $(-2) + (-5)$

23) $5 - (-6)$

25) $(-6) + 3$

27) $4 - 7$

29) $(-7) + 7$

2) $4 - (-1)$

4) $(-6) + 8$

6) $(-8) - (-3)$

8) $7 - 7$

10) $(-4) + (-1)$

12) $(-1) + (-6)$

14) $(-8) + (-1)$

16) $(-1) - 8$

18) $5 - 7$

20) $(-5) + 7$

22) $1 + (-1)$

24) $8 - (-1)$

26) $(-3) + (-1)$

28) $7 - 3$

30) $(-3) + (-5)$

Find each product.

31) $(4)(-1)$

33) $(10)(-8)$

35) $(-4)(-2)$

37) $(-7)(8)$

39) $(9)(-4)$

41) $(-5)(2)$

43) $(-5)(4)$

45) $(4)(-6)$

32) $(7)(-5)$

34) $(-7)(-2)$

36) $(-6)(-1)$

38) $(6)(-1)$

40) $(-9)(-7)$

42) $(-2)(-2)$

44) $(-3)(-9)$

Find each quotient.

46) $\frac{30}{-10}$

48) $\frac{-12}{-4}$

50) $\frac{30}{6}$

52) $\frac{27}{3}$

54) $\frac{80}{-8}$

56) $\frac{50}{5}$

58) $\frac{48}{8}$

60) $\frac{54}{-6}$

47) $\frac{-49}{-7}$

49) $\frac{-2}{-1}$

51) $\frac{20}{10}$

53) $\frac{-35}{-5}$

55) $\frac{-8}{-2}$

57) $\frac{-16}{2}$

59) $\frac{60}{-10}$

A.2. Fractions

Objective: Simplify, add, subtract, multiply, and divide with fractions

Working with fractions is a very important foundation to algebra. Here we will briefly review simplifying, multiplying, dividing, adding, and subtracting fractions. As this is a review, concepts will not be explained in detail as other lessons are.

World View Note: The earliest known use of fraction comes from the Middle Kingdom of Egypt around 2000 BC!

Simplifying Fractions

We always like our final answers when working with fractions to be simplified. Simplifying fractions is done by dividing both the numerator and denominator by the same number. This is shown in the following example.

Problem 1 : | *Worked Example*

Simplify: $\dfrac{36}{84}$

Solution.

$\dfrac{36}{84} = \dfrac{9 \cdot 4}{21 \cdot 4}$ 　Both numerator and denominator have common factor of 4

　　　　simplify

$\dfrac{9}{21} = \dfrac{3 \cdot 3}{3 \cdot 7}$ 　Both numerator and denominator have common factor of 3

　　　　simplify

$= \dfrac{3}{7}$ 　Our Solution

The previous example could have been done in one step by factoring 12 from both numerator and denominator. It is not important which method we use as long as we continue simplifying our fraction until it cannot be simplified any further.

Multiplying Fractions

We can multiply fractions by multiplying straight across, multiplying numerators together and denominators together.

Problem 2 :	Worked Example

Multiply: $\dfrac{6}{7} \cdot \dfrac{3}{5}$. Be sure to write answer in simplest form.

Solution.

$$\dfrac{6}{7} \cdot \dfrac{3}{5} \qquad \text{Multiply numerators across and denominators across}$$

$$= \dfrac{18}{35} \qquad \text{Our Solution}$$

When multiplying, we can simplify fractions before we multiply. We can either simplify vertically or diagonally, as long as we use one number from the numerator and one number from the denominator.

Problem 3 :	Worked Example

Multiply: $\dfrac{25}{24} \cdot \dfrac{32}{55}$. Be sure to write answer in simplest form.

Solution.

$$\dfrac{25}{24} \cdot \dfrac{32}{55} \qquad \text{25 and 55 share a factor of 5 while 32 and 24 share a factor of 8}$$
$$\text{simplify}$$

$$= \dfrac{5}{3} \cdot \dfrac{4}{11} \qquad \text{Multiply numerators across and denominators across}$$

$$= \dfrac{20}{33} \qquad \text{Our Solution}$$

Dividing Fractions

Dividing fractions is very similar to multiplying with one extra step. Dividing fractions requires us to first take the reciprocal of the divisor, change the division sign to multiplication and then proceed just as we do with multiplication of fractions.

Problem 4 :	Worked Example

Divide: $\dfrac{21}{16} \div \dfrac{28}{6}$. Be sure to write answer in simplest form.

Solution.

$$\dfrac{21}{16} \div \dfrac{28}{6} \qquad \text{Multiply by the reciprocal}$$

$$= \dfrac{21}{16} \cdot \dfrac{6}{28} \qquad \text{21 and 28 share a factor of 7 while 6 and 16 share a factor of 2; simplify}$$

$$= \dfrac{3}{8} \cdot \dfrac{3}{4} \qquad \text{Multiply numerators across and denominators across}$$

$$= \frac{9}{32} \qquad \text{Our Solution}$$

Problem 5 : | *Media/Class Example*

Perform the indicated operation. Be sure to write the answer in simplest form.

a) $\frac{4}{9} \div \frac{16}{27}$

c) $\frac{10}{9} \div 6$

b) $\frac{1}{6} \cdot -\frac{3}{5}$

d) $-\frac{3}{7} \cdot \frac{21}{15}$

Problem 6 : | *You Try*

Perform the indicated operation. Be sure to write the answer in simplest form.

a) $\frac{6}{11} \cdot \frac{22}{15}$

c) $\frac{8}{7} \div 14$

b) $\frac{1}{3} \div -\frac{8}{3}$

d) $-\frac{2}{5} \cdot \frac{25}{8}$

Adding and Subtracting Fractions with Common Denominator

To add and subtract fractions, we must have a common denominator. If the fractions already have a common denominator, we just add or subtract the numerators and keep the denominator.

Problem 7 : | *Worked Example*

Add: $\dfrac{7}{8} + \dfrac{3}{8}$. Be sure to write answer in simplest form.

Solution.

$$\dfrac{7}{8} + \dfrac{3}{8} \qquad \text{Same denominator, add the numerators } 7+3$$

$$= \dfrac{10}{8} \qquad \text{Simplify answer by dividing numerator and denominator by 2}$$

$$= \dfrac{5}{4} \qquad \text{Our Solution}$$

While $\dfrac{5}{4}$ can be written as the mixed number $1\frac{1}{4}$, in algebra we generally use the improper fraction instead of the mixed number.

Problem 8 : | *Worked Example*

Subtract: $\dfrac{13}{6} - \dfrac{9}{6}$. Be sure to write answer in simplest form.

Solution.

$$\dfrac{13}{6} - \dfrac{9}{6} \qquad \text{Same denominator, subtract the numerators } 13-9$$

$$= \dfrac{4}{6} \qquad \text{Simplify answer by dividing numerator and denominator by 2}$$

$$= \dfrac{2}{3} \qquad \text{Our Solution}$$

Adding and Subtracting Fractions with Unlike Denominators

To add or subtract fractions with unlike denominators, we must first find the least common denominator (LCD).

Problem 9 : | *Worked Example*

Add: $\dfrac{3}{8} + \dfrac{11}{12}$. Be sure to write answer in simplest form.

Solution.

If the denominators do not match, we will need to identify the LCD, and then build up each fraction by multiplying the numerator and denominator by the same number.

$$\dfrac{3}{8} + \dfrac{11}{12} \qquad \text{LCD is 24}$$

$$= \dfrac{3}{3} \cdot \dfrac{3}{8} + \dfrac{11}{12} \cdot \dfrac{2}{2} \qquad \text{Multiply first fraction by } \dfrac{3}{3} \text{ and the second by } \dfrac{2}{2}$$

$$= \frac{9}{24} + \frac{22}{24} \qquad \text{Same denominator, add numerators}$$

$$= \frac{31}{24} \qquad \text{Our Solution}$$

Problem 10 : | *Media/Class Example*

Perform the indicated operation. Be sure to write the answer in simplest form.

a) $\dfrac{6}{5} - \dfrac{8}{5}$

c) $\dfrac{4}{7} + \dfrac{1}{3}$

b) $\dfrac{5}{6} + \dfrac{2}{15}$

d) $-\dfrac{1}{3} - \left(-\dfrac{8}{5}\right)$

Problem 11 : | *You Try*

Perform the indicated operation. Be sure to write the answer in simplest form.

a) $\dfrac{4}{7} + \dfrac{6}{7}$

c) $\dfrac{3}{5} + \dfrac{1}{3}$

b) $\dfrac{5}{9} - \dfrac{1}{6}$

d) $-\dfrac{1}{2} - \left(-\dfrac{2}{3}\right)$

Practice: *Fractions*

Simplify each. Leave your answer as an improper fraction.

1. $\dfrac{42}{12}$

2. $\dfrac{35}{25}$

3. $\dfrac{54}{36}$

4. $\dfrac{45}{36}$

5. $\dfrac{27}{18}$

6. $\dfrac{40}{16}$

7. $\dfrac{63}{18}$

8. $\dfrac{80}{60}$

9. $\dfrac{72}{60}$

10. $\dfrac{36}{24}$

Perform the indicated operation and simplify.

11. $\dfrac{-2}{9} \div \dfrac{-3}{2}$

12. $(9)\left(\dfrac{8}{9}\right)$

13. $\dfrac{2}{5} + \dfrac{5}{4}$

14. $\dfrac{3}{2} - \dfrac{15}{8}$

15. $\left(-\dfrac{2}{9}\right)(2)$

16. $-2 \div \dfrac{7}{4}$

17. $\dfrac{9}{8} + \left(-\dfrac{2}{7}\right)$

18. $\dfrac{1}{10} \div \dfrac{3}{2}$

19. $\left(\dfrac{13}{8}\right)(-2)$

20. $1 + \left(-\dfrac{1}{3}\right)$

21. $-\dfrac{9}{7} \div \dfrac{1}{5}$

22. $\dfrac{3}{7} - \dfrac{1}{7}$

23. $\dfrac{8}{9} \div \dfrac{1}{5}$

24. $\left(-\dfrac{1}{2}\right) + \dfrac{3}{2}$

25. $(-\dfrac{6}{5})(-\dfrac{11}{8})$

26. $(2)(\dfrac{3}{2})$

27. $-\dfrac{1}{9} \div -\dfrac{1}{2}$

28. $\dfrac{5}{3} - \left(-\dfrac{1}{3}\right)$

29. $\dfrac{1}{5} + \dfrac{3}{4}$

30. $\dfrac{-3}{2} \div \dfrac{13}{7}$

31. $\dfrac{11}{6} + \dfrac{7}{6}$

32. $\left(\dfrac{2}{3}\right)\left(\dfrac{3}{4}\right)$

33. $6 - \dfrac{8}{7}$

34. $\left(-\dfrac{15}{8}\right) + \dfrac{5}{3}$

35. $\dfrac{1}{2}\left(-\dfrac{7}{5}\right)$

36. $\dfrac{3}{5} + \dfrac{5}{4}$

37. $-1 \div \dfrac{2}{3}$

38. $\dfrac{1}{3} + \left(-\dfrac{4}{3} \right)$

40. $(8)\left(\dfrac{1}{2} \right)$

39. $(-1) - \left(-\dfrac{1}{6} \right)$

41. $-\dfrac{5}{7} - \dfrac{15}{8}$

A.3. STORY PROBLEMS

Objective: Solve application problems by creating and solving a linear equation.
Problem Solving Strategies and Tools (PSST)
When first looking at an application problem (or story problem), it is often helpful to read the entire problem and then read it again more slowly to organize your thought, and note if additional information is needed.

A) Identify the unknown quantity.

B) Write an equation or inequality that models the relationship between the known and unknown quantities.

C) Solve the equation or inequality. Check for reasonableness of solution.

D) Report the solution.

| **Problem 1 :** | *Worked Example - Building a Patio* |

Olive has an area in her backyard which is 9 feet long by 5 feet wide. She would like to dig up the grass and replace it with brick. Each brick is 8 inches long and 4 inches wide.

a) How many bricks will Olive need for the project?

b) If bricks cost $0.44 each and sales tax is 9.5%, how much will the bricks, including tax, cost?

Solution.

a)

$12 \text{ inches} = 1 \text{ foot}$ Additional information needed

$9 \text{ feet} \cdot \dfrac{12 \text{ inches}}{1 \text{ foot}} = 108 \text{ inches}$ Convert backyard length from feet to inches

$5 \text{ feet} \cdot \dfrac{12 \text{ inches}}{1 \text{ foot}} = 60 \text{ inches}$ Convert backyard width from feet to inches

$A = lw$ Area of a rectangle
$A = 108 \text{ inches} \cdot 60 \text{ inches}$ Find area of the backyard
$A = 6480 \text{ square inches}$

$A = 8 \text{ inches} \cdot 4 \text{ inches}$ Find area of each brick
$A = 32 \text{ square inches}$

$\dfrac{6480}{32} = 202.5$ $\dfrac{\text{Backyard Area}}{\text{Area of Brick}} = \text{number of bricks needed}$

Olive will need 203 bricks for her backyard.

b)

$$(203\,\text{Bricks})\left(\frac{\$0.44}{\text{Brick}}\right) = \$89.32 \quad \text{Cost of Bricks without tax}$$

$$(0.095)(\$89.32) = \$8.49 \quad \text{Tax}$$

$$\$89.32 + \$8.49 = \$97.81 \quad \text{Cost of Bricks} + \text{Tax}$$

Total cost of bricks, including tax is $97.81.

Problem 2 : | *Media/Class Example - Gardening*

Peter wants to grow more pumpkins in his small backyard garden. The current dimensions of his garden are 5 feet long and 4 feet wide. He wants to increase the perimeter by 40%.

  a) What are the new dimensions of his garden?

  b) What is the new area? Round your answer to one decimal place.

  c) By what percent did the area increase?

Problem 3 : | *Media/Class Example - Backyard Sandbox*

Sammy was given a sandbox for his birthday but it is not yet filled with sand. The sandbox is 4 square feet and 12 inches deep.

  a) How many cubic feet of sand is needed to fill Sammy's sandbox?

  b) Fifty pounds of sand is 0.5 cubic feet and cost S4.15. How many bags are needed?

| Problem 4 : | *You Try - At the Grocery Store* |

When checking the weekly grocery advertisement, Paul notices 13.5 ounces of canned spinach is on sale, 3 cans for $9.98. What is the price for 20 cans?

| Problem 5 : | *You Try - Barbecue Grill* |

Wilbur needs a new barbecue grill and finds one on sale for 20% off plus a $25.00 mail in rebate. The original cost of the grill is $369.00. What is the discount and what amount will Wilbur pay for the grill?

Practice: *Story Problems*

Solve the following story problems. Be sure to report your solution in a complete sentence.

1. *Halloween:* Based on previous years, Sheldon estimates the number of trick or treaters who will come to his door this year will be 92. His treat supply includes 150 assorted chocolate bars and 72 bags of assorted chips. How many treats will Sheldon be able to give to each little ghost or goblin and how many will be left-over?

2. *Halloween:* The 157 million Americans planning to celebrate Halloween will spend an estimated:

Decorations	$1.9 Billion
Candy	$2.1 Billion
Costumes for Pets	$550 Million

Source: National Retail Federation (October 25, 2015 Parade Magazine)

What is the total amount of sales?

3. *Holiday Baking:* Sabrina has a recipe for a full size cake but the pumpkin shaped pan she wants to use is $\frac{2}{3}$ the volume needed for the recipe. The recipe calls for $2\frac{1}{4}$ cups of flour. How much flour will she use?

4. *Baking:* Kimmie's recipe for fudgy brownies calls for 1 lb. of chocolate chips. His kitchen scale is not accurate so he needs to measure the chocolate chips from his giant $4\frac{1}{2}$ lb. bag. After some online research, Kimmie finds that 2 cups of chocolate chips is 12 ounces. How many cups of chocolate chips will she need for this recipe? Write your answer as a mixed number.

5. *At the Coffee Stand:* Before going to class, Amy stops at the coffee stand to buy 12 ounce cup of hot chocolate and a raspberry scone. The hot chocolate is $3.25 and the scone is $2.95 (tax is included in the purchase price). Amy gives the barista $10.25. How much change will she get back?

6. *At the Coffee Stand:* After her midterm exam, Jasmine and two friends go to the coffee stand for post exam treats. They each order a tall cappuccino and share a large snicker doodle. The drinks are $3.85 each and the cookie is $2.55. What is the total amount due and how would they split the bill equally among the three friends?

7. *At the Grocery Store:* Belle stops at the grocery store to pick up a few items and realizes she left her debit card at home and has just $12.27 in cash. Will she have enough cash to buy a $2.49 bag of tortilla chips, one jar of salsa that is two for $5.39 amd four soft drinks at $1.29 each. If Belle has enough cash, how much does she have leftover? If she does not have enough cash, how much more is needed?

8. *Ordering Pizza:* Raphael is ordering pepperoni pizza and would like the better deal. Deep dish pizza is 1 inch deep and regular crust is $\frac{3}{8}$ inch deep. Find which size costs less per cubic inch: a 12-inch deep dish pizza for $21.50 or a regular crust 15-inch pizza for $20.95.

9. *Making Lasagna:* Mark and Jill are making lasagna from ingredients found onlin and delivered to them. They will order two 9-ounce boxes of lasagna for $1.78 each, 4 jars of spicy marinara and roasted garlic sauce for $18.14 and four packages of finely shredded mozzarella cheese for $15.39. How much will the lasagna cost if they need both boxes of lasagna, 2 jars of marinara sauce and three packages of mozzarella cheese?

10. *Baseball Pizza:* The day after the Seattle Mariners make a double play, pizza is half price. The regular price for a large one topping pizza is $19.00. Extra toppings are $1.50 each. Michael orders one Hawaiian pizza (two toppings) and one mushroom, black olive and onion pizza plus a $3.75 container of gelato. Delivery is $3.00 and tax is included in the menu price. How much is the total bill?

11. *Sidewalk Lemonade Stand:* Louie is selling lemonade for $1.25 a cup. The cost to make one cup of lemonade is $0.27 amd each plastic cup is $0.16. If he sells 22 cups of lemonade, will he make at least $18.75, enough for a movie ticket and snack?

12. *Boating:* Joe and Amanda are going to Greenlake for an afternoon of boating. The cost to rent a pedal boat is $18.00 an hour or $6.00 for 15 minutes. After their one and a half hour pedal boat ride, they each buy a large iced cappuccino ($3.75 each). How much money did they spend at Greenlake?

Body Mass: Body Mass Index (BMI) is a person's weight in kilograms divided by the square of height in meters. If a person has a BMI between 18.5 and 24.9, that person is considered to have a healthy weight.

13. *Healthy Weight:* Leonard is 5 feet 5 inches tall and weighs 148 pounds.

   a) What is Leonard's BMI? Round to two decimal places.

   b) Is it in the healthy range? If not, how much weight should he gain or lose to be in the healthy range?

14. *Healthy Weight:* Patsy is 4 feet 5 inches tall and weighs 158 pounds.

   a) What is Patsy's BMI? Round to two decimal places.

   b) Is it in the healthy range? If not, how much weight should she gain or lose to be in the healthy range?

Made in the USA
San Bernardino, CA
07 September 2018